T0360813

Community Real Estate Development

Community Real Estate Development: A History and How-To for Practitioners, Academics, and Students introduces the fundamentals of affordable housing to aspiring development professionals. From understanding the history informing today's affordable housing programs to securing financing and partnering with public and private stakeholders, this primer equips students and emerging professionals for success in a unique area of the real estate industry. Topical chapters written by nationally recognized leaders in community real estate development (CRED) take a didactic approach, using real-life examples and case studies to provide context for reflection. Drawing on the authors' experience as private sector developers, state and municipal housing officials, and not-for-profit executives, this versatile resource offers an insider's perspective on creating and maintaining affordable housing in any real estate market.

Features:

- Covers topics including community design, development policy, tax credits, land use planning, development rights, historic buildings, adaptive reuse, tax increment financing, and gentrification
- Presents interviews with development professionals in asset and property management, commercial real estate brokerage, and local housing authorities and government agencies
- Highlights winning case studies from a student competition to inspire similar classroom activities
- Includes a glossary of CRED-specific terminology to help readers master the language of affordable housing
- Contains diverse examples, planning tools, and "programs to make numbers work," with a companion website available

Blending the latest academic research with hard-won insights from the field, *Community Real Estate Development* prepares the next generation of affordable housing professionals to continue the work of its pioneering authors and editors.

Stephen Buckman is an Associate Professor of Real Estate Development at Clemson University. His research is centered on resiliency, waterfront development, and community real estate development. Dr. Buckman co-taught the CRED program for three classes and was a member of the CRED Steering Committee. Dr. Buckman is also a partner in a real estate development company that specializes in small community-oriented multi-family projects as well as consulting to developers and communities.

Jeff Burton manages the City of Tampa, Florida Community Redevelopment Agency. He also earned his Ph.D. from the University of South Florida, writing a dissertation focused on sustainable redevelopment. Jeff is an experienced State of Florida licensed residential building contractor and has performed over 5,000 building safety inspections as a licensed building code administrator. While at the Insurance Institute for Building and Home Safety (IBHS), he co-authored the Louisiana State building code law in the aftermath of Hurricane Katrina and was a founding research member of the State of Florida post Hurricane Charley building code analysis team. Jeff is also the President of the Florida Redevelopment Association (FRA).

John Talmage is Director of the Lee County Office of Economic Development. He is responsible for attracting $5 billion of new development to Lee County ranging from affordable housing to corporation headquarter attraction to new community revitalization initiatives.

Community Real Estate Development

A History and How-To for Practitioners, Academics, and Students

Edited by
Stephen Buckman
Jeff Burton
John Talmage

Routledge
Taylor & Francis Group

NEW YORK AND LONDON

Cover image: Jeff Burton

First published 2023
by Routledge
605 Third Avenue, New York, NY 10158

and by Routledge
4 Park Square, Milton Park, Abingdon, Oxon, OX14 4RN

Routledge is an imprint of the Taylor & Francis Group, an informa business

Library of Congress Cataloging-in-Publication Data
A catalog record for this title has been requested

ISBN: 978-0-367-62555-9 (hbk)
ISBN: 978-0-367-62554-2 (pbk)
ISBN: 978-1-003-10967-9 (ebk)

DOI: 10.1201/9781003109679

Typeset in Times New Roman
by codeMantra

Access the companion website: communityredev.com

Contents

CRED Contributors List

Editors:

Stephen Buckman (Chs 1, 3, 8, 14)
Jeff Burton (Chs 1, 2, 7, 14)
John Talmage (Chs 1, 14)

Contributors:

M. Martin Bosman (Ch 8)
Angela Crist (Foreword)
Trent Green (Ch 6)
Soomin Kim (2018 Winner)
Evangeline Linkous (Ch 4)
Camille Reynolds Lewis (2019 Winner)
Barry Stiefel (Ch 5)
Elizabeth Strom (Ch 2)

Foreword

Community Development has always been plagued with a lack of consensus to definition. Does It empower communities and individuals, encourage them to adopt sustainable goals, organize external resources to meet the needs of an existing community, some of these, or all of these? There are themes that connect all of these goals, but is it community development? Community development simply is not a term of art. *Community Real Estate Development* seems like a simpler term to define since it is limited to real estate development, but is it any real estate transaction – market-based apartments, upscale grocery, or a high-end coffee shop or does where it is located and who it serves become what is important? While this book does not address directly what the definition of *Community Real Estate Development* is, it does go into detail what are the different components of the development process that a community development agency or another practitioner needs to follow. Putting together a team, identifying a site, conducting a market analysis, building pro formas, and obtaining financing are just some of the steps that a developer will have to master to move the idea of a development into a buildable project. Along the way, the developer should be asking how does this benefit the community, do I have their support, what other type of investment could this leverage, and does it serve the existing community or a new one that might move it. These begin to lay the foundation if it is a community development project or not.

Community development, as a field of practice, has followed a long circuitous route to where it is now. In the late 19th century, Community Development saw the convergence of Agricultural Extension Service with urban public hygiene and fire safety regulations. It grew out of the extension services since most Americans lived in rural America. Further, after the great fires that beset our cities at the turn of the 20th century, addressing overcrowding and design malfunctions became a public priority to address. As the century progressed and community development became more institutionalized, it took on more of an economic and housing focus. As the federal government stepped in to fund community development projects, it moved into an urban renewal tool to remove what was considered blighted communities that had long been defined by substandard tenement housing.

This was followed by large infrastructure projects such as the development of the interstate highway system which displaced hundreds of thousands of residents from low-income urban communities. Two issues associated with Federal participation was the federal government's proclivity to make decision centrally where the community development model is making decisions locally so that they reflect local learnings and second, the community development field was not able to absorb the amount of funding that was put into the system in a meaningful way. This led to the overuse of planners and grant writers who often had neither community development nor planning experience. Today, there are new forces influencing the direction of community development shifting its focus from subsiding housing to often the privatization of the same housing. There is a bias to find market-based solutions and depend on public-private partnerships.

It is the authors' hope that publishing this book will begin to build on a community of users at the grassroots who are interested in being advocates for their community, as well as investors and builders. For the past ten years, the class, which this book is based upon and inspired by, has been produced at the University of South Florida's (USF) Florida Institute of Government with strong support from the Hillsborough and Pinellas Housing Finance Authority and the Tampa Housing Authority. This class was originally developed by the late Thomas Zuniga, in a class, he had developed for LISC-South Florida. Once their office closed, Tom contacted Ed Johnson, the executive Director of the East Tampa CRA who knew and worked with Tom in a variety of positions through the years, but easily understood that if a class like this could be made available in the Tampa Bay area, then he had a responsibility to do it. Ed worked with several departments throughout the University, finally finding a more permanent home at USF's FIOG. The class has had a number of facilitators since Tom's passing and has brought the class to Palm Beach, FL, Broward County, FL, and Jacksonville, FL. To date, the class has graduated approximately 500 alumni working in every area of community development and beyond – finance, real estate, government, federal regulatory agencies, banking, and the list could go on and on.

What the book gives is the flavor of and the strength of the class as well as the diversity of presenters, students, and organizations that supported the program. The blend of graduate students from USF and students who were working in community development as were practitioners was a special experience in and of itself. Students were faced with new skill sets and points of view that often took them out of their comfort zone to challenge them to think about solutions in a different way. The course was long enough – six weeks, for a full day and a half per week – where participants got to know one another well enough to get beyond some of the courtesies of typical classroom gatherings. The class also included working on a team project that met every week during the course to put together a project that was judged by industry experts in a competition to select the best project. Often the winning projects would go on to become funded projects and the class can point with pride to any number of affordable housing developments, community gardens, and incubators developed around the state that were envisioned and shaped here.

While there is more work to be done – perhaps a CRED2.0, this book is the capstone to the efforts of the FIOG to support *Community Real Estate Development* in the State of Florida and make it accessible to all communities who wish to move a community's vision to blueprints, construction documents, and a final built product.

Angela Crist, Director Florida Institute of Governments

Contributors

M. Martin Bosman: Dr. Bosman is an Associate Professor of Geography at the University of South Florida. His research focuses on global city formations, the political economy and ecology of urbanization, and the politics of place-competition. He is particularly interested in the politics of anti-urban economic development and the rift in the metabolic relations between cities and nature.

Trent Green: Dr. Green is an Associate Professor of Architecture and Urban Design at the University of South Florida's School of Architecture & Community Design. Dr. Green's areas of research include urban/community design planning, economic development/community revitalization, housing/residential development strategies, and development regulations.

Evangeline Linkous: Dr. Evangeline Linkous is the Program Director for and an Associate Professor in the Master of Urban and Regional Planning program at the University of South Florida. She studies the fiscal and legal frameworks used to rationalize management of land resources. Her research has been published in journals including the *Journal of the American Planning Association, Land Use Policy, Urban Affairs Review*, and the *Journal of Environmental Planning and Management*. She has a Ph.D. and Master of Urban and Regional Planning degree from the University of Pennsylvania, and a Bachelor of Arts in English from New College of Florida.

Barry L. Stiefel: Dr. Stiefel is an Associate Professor in the Historic Preservation and Community Planning program at the College of Charleston. He is interested in how the sum of local preservation efforts affects regional, national, and multi-national policies within the field of cultural resource management, natural heritage conservation, and land use planning. Dr. Stiefel has published numerous books and articles.

Elizabeth Strom: Dr. Strom is an Associate Professor in the School of Public Affairs at the University of South Florida. She previously served as the Director of USF's Office of Community Engagement and Partnerships. Her research interests include housing policy and tourism and how it shapes our cities.

The Background of CRED

1 Introduction to Community Real Estate Development

Stephen Buckman, Jeff Burton, and John Talmage

Critical Directions and a CRED Overview

The present state of the real estate industry over the last few years is one of the rapid increases in terms of infrastructure and finances. Entire communities have been precipitously changed in some cases for the better and in some for the worse, depending on which side of the gentrification aisle you sit on. What is being lost in the rapid real estate development expansion is the community aspect.

The term "community" represents a very broad swath of what it entails. Thus, when discussing community real estate development (CRED) which is the thrust of this book, community should be looked at in a multitude of ways. It is designed to be looked at and considered in many formats when discussing it and coupling it with real estate development. In one regard, the community aspect of real estate development can be seen in terms of affordable housing which helps make sure all members of the community are able to have a decent and just set of housing options. Yet, it can also entail a more direct link to community reshaping in terms of development concepts that include aspects such as an e-gaming after school center for neighborhood youth to co-op grocery stores and artist housing and studios.

Thus, CRED needs to be seen as a form of real estate development that is not entirely concerned with the bottom line and return on investment (ROI) but rather is also concerned with how a development will impact the surrounding community and population. It is also about developers and developments that are invested in the neighborhoods in which they choose to develop. As one of our interviewees, Joe Bonora pointed out, what separates CRED from standard market rate developments is that CRED developers are in it for the long game. The time horizons for a market rate development of two–five years (often 10 at the most), which is common, pale in comparison to the CRED developer who is looking most often at time horizons of 10–15 years. Thus, CRED is a long game plan.

Being part of the surrounding neighborhood and often being built in areas that are not on the radar of other developers because of the socio-economic make-up of the community, the CRED developer will often help

DOI: 10.1201/9781003109679-2

facilitate change in an area that increases its economic and social presence. Yet, this resurgence of a neighborhood takes time, in turn increasing the time horizon of a successful development. So, CRED sees a successful development as being one that helps strengthen the community and, in the process, instills a sufficient ROI for its investors. As the development in question will often help re-establish a neighborhood, that re-establishment will help increase the ROI of the development as the value of the building(s) will increase as the area around grows and prospers, thus increasing the valuation of the asset. But this valuation is a result of a build and hold strategy that takes the long game into account.

So, as one can see, by no means should it be looked at that CRED is not about making a profit for the developer and their investors. But what separates this form of development from a market rate development is that while the ROI is a key variable, it is not the only variable to be thought about.

CRED Class

The impetus for this book and for the ideas presented here come from a class that has been co-developed and sponsored by the Tampa Housing Authority and Florida Institute of Governments and based out of the School of Public Affairs at the University of South Florida (USF). The six weekend class is composed of roughly 20 students: generally, half of grad students are from USF and half from the community taught students, what could be considered a standard suite of real estate development 101 classes that would be learned over a year in a university-level real estate development program. These classes covered the gamut of real estate development which included: market analysis, site analysis, development proposals, zoning, public speaking, and proforma building.

While the idea behind the CRED class is for it to be introductory, it was also intended to be as "real world" as possible. From the outset of the class, the "real" component of the class was initiated. One of the first things the students do is to pitch an idea for development that then is voted on by the class, with the top three or four ideas moving forward as "real world" development examples. From there, the students are split into groups and assigned to one of the three development ideas. From that point, all of the class instruction is geared around their particular development.

The faculty and guest speakers work with each team to shape their development to not only make a profit but also have a community impact. The faculty sees themselves as guides throughout the process. At the end of the six weekend class, each of the teams presents their final development proposal. The presentation in and of itself is a real world activity in that the students present their proposals to a board that consists of various members of the community that include but are not limited to developers, bankers, and equity brokers who acknowledge they will work with the winning team to help them turn their development proposal into reality.

Outline of the Book

This book is composed of two primary sections. Section 1 will constitute an overview of the field of CRED. Section 2 consists of stakeholder interviews in the CRED community (aka "voices from the field") as well as an Appendix which composes four case studies of the past winners of the CRED certification competition, key terms, and a list of government programs and agencies that augment the CRED process.

Section 1

This section is composed of nine chapters which include the introduction chapter you are reading. In these chapters, the authors lay the foundation of what CRED is and its importance, how to develop CRED, and how we got to the present stage with community and real estate development. The chapters run the gamut from the history of community development to finance to the impacts of gentrification on our built form. Section 1 is designed to give the reader the foundation of understanding the CRED world. Chapters in the section are as follows:

Chapter 2: History of Community Development by John Talmage

Chapter 2 examines the history of United States community development. It delves into the needs and mechanisms used to address community development primarily in the 20th and early 21st centuries. In this chapter, Mr. Talmage explains what past community development ideas looked like and how we have reached our present point that demands the CRED model.

Chapter 3: Need for Affordable Housing and US Housing Policy by Elizabeth Strom and Jeff Burton

In this chapter, the authors, Drs. Strom and Burton, follow Mr. Talmage's explanation of history of community development, take those ideas, and more precisely examine the history of housing policy in the United States in the 20th and early 21st century, highlighting government programs and their application. The chapter concludes by examining the current low to moderate income housing programs.

Chapter 4: Programs: History and How to Utilize Them in a Real Estate Development by Stephen Buckman

Chapter 4 presents an overview of key government programs used to help balance the playing field in terms of real estate development. The use of government-subsidized programs to offset the losses of doing community-oriented work is key to the success of any CRED or affordable housing driven development. In this chapter, Dr. Buckman further explores government programs used to encourage developers to build community-driven developments. The programs explored in this chapter include: Low-Income Tax Credits; Historical Tax Credits, Opportunity Zones; and New Market Tax Credits.

Chapter 5: Transfer of Development Rights and CRED by Evangeline "Van" Linkous

In Chapter 5, Dr. Linkous examines the importance and use of community development Transfer of Development Rights (TDR). Many cities look to focus on certain development uses, including community development in key locations. Often, these locations do not align with the developer's wishes. She explains how TDR allows developers to trade their site location's present developments rights for the rights to develop in another more beneficial location to the community and developer. The use of TDR can be used by cities to better situate needed community-oriented developments in areas where they are most needed and will have the greatest impact.

Chapter 6: Historical Reuse as a form of CRED by Dr. Barry Stiefel

Chapter 6 addresses historical building reuse. Many communities that were once considered a low development priority are being rediscovered for their redevelopable urban cores. In turn, many undesirable historic buildings are now in-demand development sites. Historic redevelopments have higher standards and are, at times, not practical for traditional developers. Thus, in this chapter, Dr. Stiefel examines the use of different historical development strategies incentives and credits that encourage community-driven historic rehabilitation.

Chapter 7: Community Design by Professor Trent Green

Chapter 7 examines the importance of urban design within our built forms. Community and building form are important in creating more community-driven developments. Thus, in this chapter, Professor Green looks at the impact of design in community development and presents examples that encourage more equitable communities. This chapter further highlights the policy/design relationships and how they can lead to vibrant communities and their importance related to community/lower income-driven developments.

Chapter 8: Community Redevelopment Agencies by Dr. Jeff Burton

Chapter 8 identifies state-by-state redevelopment statutes and discusses their CRED usage. These redevelopment entities can be important local community economic development mechanisms as they channel funds into and encourage projects that revitalize slum and blighted areas. Dr. Burton, in this chapter, encourages readers to learn what drives these entities (their mission and purpose) and how they may partner with CRED developers. Dr. Burton also presents financial structures that can make this happen.

Chapter 9: Balancing the Gentrification Elephant in the Room by Dr. Martin Bosman and Dr. Stephen Buckman

Looking into development's dark side, in Chapter 9. Drs. Bosman and Buckman address the "elephant in the room" gentrification. Even with the greatest of intentions, gentrification and the displacement of long-time community members are of paramount concern to CRED. Understanding the balance between development and gentrification is a foundational concern for community-driven development. Understanding a development may not directly gentrify the neighborhood but may increase the likelihood that a community becomes more enticing for market-driven development. This chapter will examine gentrification's from three key perspectives (class, race, and finance) and what can be done to combat gentrification.

Section 2

In Section 1, we presented CRED themes of what came before and what measures could be used to make CRED a reality. In Section 2, we examine CRED in action. In this section, we combine interviews of the players in the field who are engaged in day-to-day CRED related activities with exposés of past winners and key CRED terms. We also offer a chapter on what the future of CRED and more equitable real estate development holds.

In the first part of this section, we interview people working within the CRED arena to see how they operate the world of CRED. We call this section "Voices from the Field." First, our interviews include Joe Bonora President of Catalyst Asset Management and Catalyst Community Development and discuss what separates the CRED developer from the market rate developer. We follow that with an interview with Leroy Moore who is currently the Senior Vice President/Chief Operating Officer of the Tampa Housing Authority and has worked in real estate development since 1984. We complement our interview with Mr. Moore with an interview of Lisa Reeves who is a commercial broker and a subject matter expert on the commercial real estate market in the Greater Tampa Bay Area and explains how the CRED commercial real estate market differs from standard commercial real estate.

Following the interviews with Ms. Reeves and Mrs. Moore and Bonora, we discuss how incentives such as LIHTC are being used in the day-to-day development world. In this capacity, Reed Jones and Ralph Settle of Beacon Development explain how they used tax incentives to build affordable housing which is CRED in action. Last, echoing all of the previous interviews, we interview Roxanne Amoroso who is the lead Principal at the Mosiac

Development Company and who discusses her years of community multi-family real estate development.

Supplementing the entire book is the appendix. In the appendix, we offer a list of key terms and resources where further information can be found on some of the incentives that were discussed in the books such as LIHTC and Opportunity Zones to name two. But a key aspect of the appendix is a list and brief overview of each of the four past (2018–2021) winners of the CRED development proposals.

Concluding Remarks

We intend for this book to encourage readers to think about real estate in different terms. We want them to see that real estate development does not need to be solely concerned with the ROI but also with how a development impacts the surrounding community. We hope that the lessons learned in this book can help to educate real estate students, developers, and housing authority officials understand how real estate development can benefit the community as well as the financial bottome line.

At the end of the day, as our cities and built forms become more gentrified due to real estate valuation, it is not up to government policy but the real estate development profession to be built better. It is our belief and it is also yours after reading this book that developers have an ethical if not moral responsibility to make our communities a better place. The development community in the end has the greatest impact on the built form as it is their skin in the game and in turn their visions that shape our built form.

2 Community Development Policy and Community Real Estate Development

Jeff Burton and Elizabeth Strom

Introduction

Community real estate development – e.g. development that serves the public beyond the profit interests of the developer doesn't exist in isolation. Rather, it is closely tied to a range of public policies that fall under the umbrella of the area we call "community development," defined by Green and Haines as "a planned effort to build assets that increase the capacity of residents to improve their quality of life" (2016, p. xiii). Policies in this area may not be easy to identify, because often they are not "labeled" community development – rather, programs focused on building housing; on environmental sustainability; on generating employment in marginalized areas are all part of the community development arena.

In this chapter, an overview of public policies that have shaped community development and community real estate development is provided. While community development policies can be found at all levels of government, the focus of this chapter will be on federal policies, which often provide a framework for actions at other levels of government and in the private and nonprofit sectors. The number of such policies can, of course, be overwhelming. For our purposes, the focus will be on major initiatives whose goals are geographically targeted, place-based, and have the physical transformation of an area as a goal. These, after all, will have the most direct connection to community-based real estate initiatives.

General Overview

Scholars have sought to synthesize the federal approach to community development either thematically or chronologically. Roberts (2016), writing from a cross-national perspective, is able to identify key themes and eras that are as relevant to the United States as to the United Kingdom. First, communities must respond to imperatives for physical transformation, first accommodating the needs of an industrial city and more recently building for the information economy. Next, we see ever-present concerns about providing safe and accessible housing for a broad range of households, an effort

DOI: 10.1201/9781003109679-3

that was especially fraught in the early 20th century but remains present today. Planning and building for either growth or contraction (depending on the urban area) is another theme found across community development efforts, and all of this takes place amidst evolving federal/national policy arenas, and a growing concern about environmental sustainability.

Focusing directly on the US experience, Alice O'Connor puts community development into a larger policy/political context. She points out that efforts to achieve real community development in the United States have been small and piecemeal, running into heavy headwinds of broader policies that have encouraged suburbanization and investment in Sunbelt suburbs at the expense of older industrial cities. For every million dollars invested in, say, an urban affordable housing project, there has been a few billion spent on highways that aid suburbanization. Moreover, policies that target resources to struggling places are often complicated to administer; are fragmented between agencies and levels of government; and rely on fragile political alliances. Suburban homeowners can take advantage of mortgage interest deductions, for example, that involve no more than adding a number to a tax return, whereas those seeking help to build affordable housing targeted to low-income renters must go through time-consuming and complicated competitive grant applications. No wonder community development policies may struggle and are often first on the chopping block when government budget makers look to make cuts.

Von Hoffman (2012) analyzed the evolution of community development as the story of local action vs. top-down approaches such as urban renewal. He is more optimistic than O'Connor, as he documents the increasing clout and sophistication of local community development practice over the latter part of the 20th century. He credits federal programs and philanthropic efforts with providing resource pipelines to local initiatives and points to successful revitalization strategies that have been difference makers in places like the South Bronx.

These community development policy overview perspectives can help us shape the story which will be told in this chapter, identifying key community development policy eras, and for each ask the following: How did these policies address physical redevelopment challenges in marginalized places? Who were the key actors in and beneficiaries of these policies? And finally, what are the legacies of these policies relevant to current community real estate development efforts?

Seeds of Community Development

It's common to start the community development story with the New Deal – the series of policy interventions used by the Roosevelt administration as part of an effort to move the US out of the Depression. There are good reasons for this starting point, discussed in the next section, but it is well worthwhile to start earlier. One could argue that the roots of community development are found in the late 19th-early 20th century efforts to study and then address the problems of poor housing and neighborhood conditions associated with industrial cities and their working-class residents.

First is concern that there were private as well as a few government efforts to address poor housing conditions found in cities. Reformers and journalists spearheaded efforts to shed light on and improve what were called "slum" housing conditions, referencing overcrowded, poor quality housing that was the hallmark of working-class quarters around the world. While reformers like Lawrence Veiller pushed for building code changes (usually at the state or local level) that would ensure some basic safety and sanitary quality for residents, reporters like Jacob Riis used the newly popular magazine format to bring narrative and photographic details of slum conditions into middle-class homes. The work of journalists like Riis did not skimp on the lurid detail *(Look into any of these houses, everywhere the same piles of rags, of malodorous bones and musty paper, all of which the sanitary police flatter themselves they have banished to the dumps and the warehouses. Here is a "flat" of "parlor" and two pitch-dark coops called bedrooms. Truly, the bed is all there is room for.)* because the goal was to foment outrage and hence pressure toward reform. Some of the first efforts to confront slum housing conditions came in the form of studies. In 1893, U.S. Congress allocated $20,000, or the equivalent of $573,000 in 2020 (Webster, 2020), to investigate and study conditions in New York, Chicago, Philadelphia, and Baltimore; their report defined city slums as "dirty back streets, especially such streets as are inhabited by a squalid and criminal population; they are low and dangerous neighborhoods" (United States. Bureau of, Houghton, and Olmsted, 1894). The study offered no formal mitigation to address the slum issue and took no congressional remedial action (Hill, 1952). Another federal study (again without accompanying action) was commissioned in 1909 under President Theodore Roosevelt; this report identified lax building standards as a root cause of poor housing conditions, which in turn bred other undesirable outcomes, from premature death to juvenile delinquency. The committee lamented that the working poor could not afford higher rents for better living conditions (United States. President's Homes and Sternberg, 1909). This report, with a focus on Washington, DC, was unusual in that it noted the failure of the private market to address the housing needs of the poor; the committee recommended government loans for slum clearance, low-income rentals, and purchasable, habitable dwelling construction (Milgram et al., 1994).

At the state and local level, however, research did at times lead to action. The New York state legislature formed the 1898 Tenement House Committee that led to the New York Tenement House Act of 1901 (Social Welfare History, 2018). This law addressed new building requirements of outward-facing windows, indoor bathrooms, proper ventilation, and other fire safeguards and introduced chapter titles such as "light and ventilation" and "sanitary provisions" (The City of New, 1901). Historians have subsequently debated whether such building code reforms had a significant impact on the health and comfort of tenement dwellers (Hill, 1952; Schilling and Pinzón, 2016), but this law remains significant as one of the first times government powers were used to shape market forces with the goal of protecting the interests of poor city dwellers.

Tenement housing laws, which used government powers to shape residential markets in marginalized areas, represented one precursor to community development; another can be found in the voluntary efforts of the settlement house movement. Settlement houses, usually homes within immigrant neighborhoods in which upper- and middle-class volunteers (in the United States, largely women volunteers) lived and provided outreach, education, and other services to the community, were first launched in the United States in 1886. Institutions like Hull House in Chicago, or the Henry Street Settlement in New York City, conducted neighborhood needs assessments before standing up programs thought to address the most salient needs of the neighborhood. The Settlement House model has been understandably criticized by contemporary observers for failing to empower the communities in which they were situated, in fact some of the leaders of the movement did have strong advocacy goals (Wirka from Planning the 20th C American City). A few settlement houses, such as Henry Street Settlement in New York City, remained active well beyond the early 20th century and have become transformed into modern community development organizations.

In sum, the precursors of community development found in the late 19th and early 20th centuries were largely driven by middle- and upper-class reformers from the worlds of philanthropy, public health, social work, and journalism. A primary concern was using both public awareness and the powers of state and local government to create building laws that would curb the worst predations found in tenement housing.

The New Deal Brings the Federal Government to the Cities

Community development policies would undergo a sea change during the 1930s and 1940s, thanks to the interventions of a series of federal programs collectively known as the New Deal, as well as post-WWII efforts to bring the powers of the federal government to the task of urban redevelopment.

Responding to the deep dislocations associated with the Great Depression, with one-third of the workforce unemployed, Congress passed a number of pieces of legislation designed to stabilize the economy and put people back to work. The Emergency Relief and Construction Act of 1932 was designed to insert federal resources into what had previously been local relief efforts, authorizing loans to provide low-income housing or slum clearance among other ends. The Reconstruction Finance Corporation (RFC), a federal corporation established to provide emergency financing for banks, other financial institutions, and other purposes, generated the credits (Government et al., 1931). Although the primary purpose of these efforts was economic stabilization, housing and infrastructure projects were often the beneficiaries: for example, an eight-million-dollar loan (85% of the construction cost) financed New York City's Knickerbocker Village (Milgram et al., 1994), according to TIME (1934), the first experimental government-financed, low-cost housing located "smack in the middle of slum-mulligan of Manhattan's East Side." The block had been nicknamed the "lung block" due to its high tuberculosis mortality rate; it had housed 650 families and reportedly eight saloons and five brothels before its redevelopment. Interestingly, the redeveloped site was not intended to house the poorest New Yorkers: rents of $12.50 per room were considerably higher than the $5 per room that had been paid by the earlier residents (https://nyhistorywalks. wordpress.com/2012/01/11/knickerbocker-village/).

Federal community development really intensified, however, after the election of Franklin Roosevelt, whose policies were labeled by historians as a "New Deal" after a line in Roosevelt's first inaugural address. New Deal programs would reshape every aspect of urban and community development policy, from facilitating homeownership through mortgages, to building infrastructure, to investing in public art, to even engaging in large scale regional planning in the case of the Tennessee Valley Authority.

The lynchpin of the early New Deal was passage of 1933 National Industrial Recovery Act, which included the establishment of the Public Works Administration (PWA), including its Housing Division. The PWA's priority was to create jobs and "prime the pump" of the floundering economy (Ebenstein, 1938). Slum clearance and the creation of publicly-owned housing for struggling families were among the projects undertaken. Williamsburg Houses in Brooklyn represents one of the early PWA efforts to address housing needs. Twelve city blocks of what was considered slum housing were cleared to construct the 20 building, 1600-unit complex. While we may be accustomed to public housing projects devoid of interesting design, the Williamsburg Houses featured eye-catching modern architecture and landscaped spaces between buildings. Its common rooms boasted murals by prominent artists. The PWA managed to construct 58 housing projects across the country; all were segregated by race. Williamsburg Houses, for example, was a whites-only project (Schwartz, 2015).

The PWA was an economic stimulus agency that happened to build public housing. The passage of the U.S. *Housing Act of 1937* (Public Law 73-479), sometimes called the Wagner-Steagall Act after its Congressional sponsors, marks the start of federal public housing programs. It established the United States Housing Authority (USHA) which provided financial support to local public housing agencies to carry out clearance and housing construction. It also created a definition of "slums" as "any area where dwellings predominate which, by reason of dilapidation, overcrowding, faulty arrangement or design, lack of ventilation, light and sanitation facilities, or any combination of these factors, are detrimental to safety, morals, and health" (Senate, 1937). The passage of the law marked a successful coalition of housing reformers, who had come to recognize that building code reform alone would not solve the problem of substandard housing, and labor unions, who favored a bill that promised construction jobs. Legislators saw the public housing program as intended for working-class citizens, "the submerged middle-class," temporarily unemployed during the Depression (Cabanella, 2018; Stoloff, 2004), rather than the poorest Americans. It also represented the incorporation into federal policy of the concept of urban "blight," a term now associated with poor housing quality and declining real estate values, and discussed almost like a disease that could spread to other communities if not stopped (Burgess, 2008; McKenzie, 1924; Reissman, 1964; Schilling and Pinzón, 2016). Slum clearance was promoted to remove blight and save surrounding communities from decline. This way of justifying the demolition of substandard housing also served the interests of the real estate sector. Because property owners feared that public housing would steal potential tenants from the private market, legislation included an "equivalent elimination" clause that required that a unit of slum housing be demolished for every public housing unit built. This furthered the efforts of communities to eliminate substandard housing, meant that public housing did not actually expand housing opportunities, and in some cases, slum clearance efforts tightened housing markets (Richter, 1983).

The country's entry into World War II not long after the passage of the Wagner-Steagall Act meant that new slum clearance and public housing projects were largely put on hold. The issue of urban conditions reemerged after the war, when housing shortages added to the stresses on cities. New Deal era housing programs, however, opened the door to the creation of community development as a policy area, and in particular an area for federal intervention.

In turn, this was the first time that urban redevelopment and housing quality were seen as an arena for federal – as opposed to state, local, or philanthropic – engagement. With federal financial support, far more ambitious urban redevelopment schemes, including both clearance and construction, were possible. Despite this federal engagement, New Deal approaches still left a great deal of discretion to local governments, in ways that would prove problematic. Slum clearance and public housing programs would ultimately run through local governments (with federal aid) after several decisions determined that the federal government could not exercise the eminent domain powers needed to carry out these projects. The 1935 *United States v. Certain Lands of Kentucky* as well as the 1936 *New York Housing Authority v. Muller* had centered around the legal argument of the federal government's constitutional authority to take private property (eminent domain) for rehousing projects and concluded that such powers could only be exercised by state and local governments. The Circuit Court of Appeals and the lower court were concerned that government-managed slum clearance was a direct step toward state communism (Ebenstein, 1938). Local control meant that communities could decide whether to build public housing and (until more recent court cases took up the mantel of fair housing) were within their communities to place it. As a result, the public housing program had exacerbated racial and economic segregation across the country.

In addressing the physical challenges of marginalized areas, New Deal programs adopted a top-down approach, where housing for the poorest laborers is seen as an eyesore and public health problem to eliminate, without much concern for the needs and preferences of those living in those places. Although housing reformers like Catherine Bauer were active in formulating policies that led to the formalization of a public housing program, to a large degree, the approach to city revitalization was driven by real estate interests and shaped by the need to assuage the concerns of landowners and their Congressional allies that federally backed housing would never compete with private housing for buyers or renters (Judd and Swanstrom, 2004). This approach would continue for several decades before encountering a backlash that would give rise to the community development movement as we know it.

Finally, many New Deal programs, including those focused on urban housing, perpetuated patterns of racial discrimination that were baked into real estate and politics at that time. Harold Ickes, Roosevelt's Secretary of the Interior under whose direction the PWA operated, did push back against those who thought public housing should be exclusively for white families. But he acquiesced to local pressures to designated housing projects for either white or African-American residents, and PWA projects followed the principle of "neighborhood composition" in which the residency of projects would reflect the "neighborhood composition" – e.g. a project built in a white neighborhood would house only white families. In some cases, racially integrated neighborhoods were razed and replaced with fully segregated public housing projects, creating new lines of segregation where they had not previously existed. According to Rothstein (2017), New Deal public housing projects reinforced and at time exacerbated racial segregation in the cities where new housing was built.

Postwar Policies

Some of the concerns about the conditions of cities and of urban housing continued once WWII ended in 1945. There had been little in the way of new housing construction or housing renovation during the Great Depression and the War (Richter, 1983), so the postwar period brought housing shortages across urban areas.

The major piece of federal legislation addressing the urban built environment was the Housing Act of 1949, which in some ways built on its predecessor Wagner-Steagall Act with some key differences. One could argue that the 1937 Act focused on housing, and slum clearance was carried out to facilitate building better quality housing (even the limitations discussed above prevented federal investments from increasing the housing supply).

The 1949 Act, however, focused on clearing areas that had been deemed "blighted," with the creation of new public housing something of a side note.

The 1949 Housing Act included idealistic language about its purposes as well as an early reference to the area of community development:

> The general welfare and security of the Nation and the health and living standards of its people require housing production and relating community development sufficient to remedy the serious housing shortage, eliminate substandard ... housing through the clearance of slums and blighted areas and the realization as soon as feasible fo the goal of a decent home and suitable living environment for every American family.
>
> (Housing Act of 1949, Public Law 81–171,
> Preamble, section 2, 81st Congress 1949)

And in keeping with these stated goals, and to win the support of an alliance of housing advocates, Title III of this act did authorize the construction of 810,000 public housing units (although the Act did not actually authorize these funds to be spent- there would be subsequent battles over spending) with income limits on tenants and cost limits on construction to ensure that these would be no-frills units built for low-income people.

But it was Title I of the Act that became its most significant contribution to urban development – a section with the title "Slum Clearance and Community Development and Redevelopment" although the program created by this section was soon dubbed "urban renewal." This section authorized payment to local governments to purchase and clear areas deemed blighted. While at first the program was intended to allow for redevelopment of these blighted areas predominantly for housing, over time (and in subsequent new legislation) the housing focus of redevelopment faded, and all manner of urban areas seen as underperforming, which could include inexpensive housing or industrial areas, were cleared for a range of commercial project. Urban renewal, then, became the lubricant for downtown redevelopment as big city mayors sought to retain the city tax base while, increasingly, middle-class residents and the commercial activities that supported them left for the suburbs. Economic development needs, not housing reform, pushed many mayors and local business communities to pursue urban renewal (Cabanella, 2018; Farrar, 2008).

The urban renewal program encouraged local governments to undertake the redevelopment of areas deemed blighted by buying up and clearing such areas, compensating owners and relocating tenants, and selling the cleared land to developers who would, it was hoped, build housing, offices, and entertainment centers that would help city centers stave off the forces of decline. The federal government supported these planning efforts, by reimbursing the municipal government for two-thirds of their costs. It's hard to find an American city whose core was not remade through these efforts. From Boston's Government Center to Philadelphia's Society Hill to New

York's Lincoln Center, urban renewal's version of community development was largely the replacement of older working-class and industrial areas with new uses, often characterized by the modernist architecture popular at that time. That's why so many cities have core areas defined by slick glass and steel skyscrapers set in sunny plazas; as Teaford (2000) notes, cities were to be cleansed of their ugly past and reclothed in the latest modern attire. There was a large coalition of powerful stakeholders who were pleased to appear at ribbon cuttings and applaud these improvements: urban core business owners sought to rally falling civic core property values, city political leaders hoped to increase their tax base, while social advocates hoped to remove slum conditions and improve poverty living conditions and enhance the housing stock. Cook (1959) noted that this law "enabled private enterprise, local government, and the federal government to combine resources in a joint attack on blight."

The urban renewal program, which ended in the early 1970s, is an important part of the community development story both for what it did and for creating a backlash that would shape urban policy for decades. Urban renewal addressed the physical challenges of declining urban areas with what became known as the "federal bulldozer" (Anderson, 1964) – clearance and complete overhaul of urban areas, often with "superblocks" replacing street grids and modernist structures replacing older forms. In some cases, these projects created public spaces and cherished structures that have been important assets to their communities. But too often, urban renewal plans did not succeed. Areas were cleared and razed, but no developers emerged interested in redeveloping the land, which in some cities sat vacant for decades.

Urban renewal played into the political and economic interests of some local stakeholders. Elected officials like Chicago's Richard Daley and agency heads like New York's Robert Moses could win accolades for new projects; corporate leaders with downtown headquarters supported developments that increased property values and created a more pleasant ambience for office workers. But those who lost homes or small businesses to clearance were often made worse off. Even when cleared land was successfully redeveloped, the urban renewal process was often difficult for those forced to relocate. Area residents were not consulted as plans were made, and efforts to help them relocate were too often inadequate. African Americans made up one-tenth of the U.S. population, while two-thirds of planned urban renewal residents (Bellush and Hausknecht, 1967), leading contemporaries to dub the program "Negro removal" (Anderson, 1964). Once displaced, tenants were often forced into more segregated and more expensive housing. Fullilove (2001) identified urban renewal as the catalyst that dispossessed African Americans of their urban land, culture, and communities and sold it to the private sector to "make way for modern developments."

In sum, urban renewal very much looked at communities through an economic lens of "blight" and sought to remove the structures and the people in them to eliminate "blight." As the construction of new public housing units never kept up with housing demand, urban renewal led to increased immiseration for those displaced, and deepening economic and racial segregation in many metro areas. Over time, the housing programs started by the 1937 and 1949 Housing Acts shifted from those primarily supported by "housers" and became a favorite of downtown "pro-growth coalitions" (Mollenkopf, 1983) who had interests in restoring property values in the urban core. The programs financed by the 1949 housing act were perfect vehicles for mayors who wished to secure their personal political futures (Judd and Swanstrom, 137). The legacy of urban renewal for community development is at best mixed. It spawned political as well as legal acceptance of the idea of using government funds and authority to remake the urban built environment, a concept that is crucial to community real estate development. On the other hand, the approach taken by many projects built through this legislation of forcing massive relocation of local residents and proceeding with almost no input from those most directly affected led to a great deal of hardship and ultimately political backlash. In some ways, urban renewal has shaped community development in that contemporary community development came into its own as an answer or correction to the processes associated with slum clearance.

Great Society and War on Poverty Programs

By the 1960s, although urban renewal efforts continued to move ahead, we saw far more organized opposition to some local plans as well as a federal government more concerned about the well-being of urban residents. Modifications to the urban renewal program would begin requiring better protection for and compensation to those forced to move, and more emphasis on improving the quality of housing in the areas to which renters were relocating, resulting in a more dramatic reframing of approaches to cities and to urban residents, most notably during Lyndon Johnson's presidency (1963–1968). Responding to urban problems in central cities that, following decades of "white flight" to the suburbs had become predominantly African American, became intertwined with an increasingly visible civil rights movement. Johnson's ambitious plans were to declare a "War on Poverty," working alongside a Democratic Congressional majority, 219 new federal programs were passed in just a few years. These included enduring social welfare initiatives such as Medicare and Medicaid, and also the creation of a new cabinet-level Department of Housing and Urban Development (HUD), set up to administer the burgeoning list of programs aimed at urban revitalization.

Perhaps the most ambitious federal urban initiative of all was the Demonstration Cities Act of 1966, which proposed coordinated housing, renewal, transportation, education, welfare, economic opportunity programs in a select number of various sized cities. The plan was to demonstrate comprehensive planning, mitigate racial discrimination in the sale or rent of housing, encourage new communities, and continue existing funding strategies. The program, which ultimately was called Model Cities, authorized the USHUD to provide grants and technical assistance to local governments to plan, rebuild, and restore entire slum and blighted neighborhoods. The gifts paid 80% of the planning and development cost, and the law authorized $24 million over two years. In addition to programs aimed at improving central city conditions, the Johnson administration also turned its attention to the racial barriers that made it difficult for people of color to move to less segregated areas. Title VIII of the 1968 Civil Rights Act established the concept of "fair housing," making it unlawful to discriminate in housing sales, rentals, or financing. It appeared that HUD was designing rules that would forge foundational changes in segregated housing trends by "opening the suburbs" to racially and economically disenfranchised groups (Bonastia, 2004).

Model Cities was one of many legislative activities during the Johnson years that could be said to give birth to the modern community development movement. First, all the Johnson era programs required what was called "maximum feasible participation" from the residents of the affected areas, and often funded organizers whose job it was to mobilize residents to take part in planning and consultation exercises. "Hence, in contrast to public housing, urban renewal, and highway construction, the antipoverty and community development projects of the 1960s enshrined, at least to some degree, a bottom-up approach" (von Hoffman, 2012 (CD) p. 7).

Even when the specific programs of the War on Poverty failed to survive the next administration, the seeding of public engagement in the decision-making process had profound effects across urban and community politics. Some of today's most established community development corporations can trace their origins to community initiatives started through these programs or to leadership initially trained through the myriad youth empowerment and grassroots education efforts.

Inner city redevelopment wasn't just the focus of federal policy makers in the 1960s; philanthropies and large corporations also sought to respond to deteriorating housing and economic conditions, especially after several summers of unrest, and these private sector programs were also crucial to the creation of the community development movement. Perhaps the best representation of how public and private sectors came together to address urban development concerns was the creation of what came to be called the Bedford Stuyvesant Restoration Corporation, perhaps the first official community development corporation in the US. Eager to have a dramatic impact in a Brooklyn, NY neighborhood that had been particularly beset by poor housing quality and unrest, Senator Robert Kennedy pushed the Johnson administration to create the Special Impact Program that would fund comprehensive local community development efforts. He also enlisted the chief executive officers of major corporations like IBM and Citibank to help raise funds and commit to employing area residents, receiving financial support from the Ford Foundation. As evidence that, at least for a brief time, aiding urban communities was a fully bipartisan effort, New York's Republican

Senator Javits and New York City Mayor Lindsay also became vocal supporters of the effort (von Hoffman, 2012 CD). With prominent young African-
American leaders at the helm, Bedford Stuyvesant Restoration could appeal
to a broad list of supporters. Civil rights leaders were enthusiastic about the
opportunities for Black-led local initiatives; liberal politicians embraced efforts to revitalize, rather than raze, a community; CEOs and conservative
leaders welcomed an initiative that stressed economic development and employment as a response to conflict (Judd and Swanstrom, 2004). Thanks to
its broad-based support, the organization is still active, having created 2,200
units or new or renovated housing as well as supported small businesses and
educational and cultural programs (https://www.restorationplaza.org/about/
history/).

The Johnson Administration, which was relatively brief (5 years), would
have an outsized significance on the creation of the community development
movement. Although top-down programs like Urban Renewal continued to
operate, the federal government took the lead in setting up a second urban redevelopment track that included grassroots mobilization and a focus on renovation rather than demolition of working-class neighborhoods. While these
War on Poverty programs have certainly received a great deal of criticism for
what they failed to accomplish (https://www.washingtonpost.com/sf/national/
2014/05/17/the-great-society-at-50/) during this period, we saw a complete rethinking of how to approach central city decline, with the primary tool being
renovation rather than demolition. The key stakeholders should be residents,
instead of or alongside elected officials, property owners, and developers.
And an important organizational vehicle for these efforts would be locally-
controlled nonprofits with community-led boards of directors, and the skills
and resources to carry out housing and economic development that meet the
needs of existing communities. These community development corporations
would remain as key actors in future community development efforts.

New Federalism and Community Development

By the 1970s, there was a shift away from ambitious federal interventions
to redevelop struggling cities. Critics, including those in the Republican
Nixon administration, had much to dislike in the alphabet soup of urban
programs produced during the Johnson years. Some questioned their effectiveness, wondering whether the millions of dollars spent were achieving
their goals of addressing poverty or revitalizing urban communities (von
Hoffman, 2012 history lessons). Others were concerned about programs
that provided funds directly to grassroots-led initiatives, bypassing governors, mayors, and local planning processes. Criticisms and scandals led
Nixon and HUD Secretary George W. Romney to announce a moratorium
on all federal building programs; Nixon also categorically unfunded future
urban renewal and threatened a veto on any new congressional spending bill
programs other than defense and budget deficit reduction (Morris, 1974).

Under Nixon, however, programs to fund community development and housing didn't end – but the design of these programs shifted dramatically, reflecting a different understanding of the appropriate role of the federal government. Rather than funding federally managed programs, rather than funding the construction and management of new housing, the federal role would be more indirect. Nixon created a series of block grants in several areas, including community development, that delivered funds to cities and states to craft their own programs within broad federal guidelines. Local officials and not HUD bureaucrats, according to this view, could best determine which park needed an upgrade, or how to meet the affordable housing needs of their residents. And while federal construction programs did resume once the moratorium was over, a new housing voucher program (originally dubbed "Section 8" because of its original authorization in Section 8 of the 1937 Housing Act), would provide rental subsidies directly to tenants who could use it to seek housing in the private market. (Hersh, 2018; Weber and O'Neill-Kohl, 2013). Interestingly, these programs have proven resilient. Community Development Block Grants continue to serve as the primary federal support for local governments, with funds distributed by a formula that considers population as well as other indicators of need, and local governments allocating funds to capital projects that meet the programs' requirements of serving low- and moderate-income households. And rental vouchers, now called Housing Choice Vouchers, continue to function as one of the primary forms of housing aid to needy families.

Another hallmark of the post-Great Society programs has been an effort to engage the private sector in the tasks of community development. This has largely taken the form of "carrots" – efforts to incentivize private investment in urban redevelopment and affordable housing efforts. The Carter administration created a program called the Urban Development Action Grant (UDAG) whose purpose was to reward cities that were successful in bringing private investment toward the task of redeveloping blighted areas. These funds spurred a generation of "public-private partnerships" toward the redevelopment of a number of many historically significant but declining central areas of older industrial cities. Now popular commercial areas such as Boston's Faneuil Hall, Baltimore's Inner Harbor, or St. Louis' Union Station area all received support from the federal UDAG program. Of course, such entertainment centers were focused more on improving government tax collections and creating jobs and less on improving conditions in marginalized communities.

The role of the private sector would be further encouraged during the 1980s, as the Reagan Administration went even further to reject federal urban intervention and focus on private sector investment as the key to revitalizing cities (R. C. Hill, 1983; Logan, 1983). Squeezed by the federal "hands off" urban policy, local public officials and community development organizations continued to pursue "public-private partnership" opportunities, leveraging public lands, planning and zoning authority, and shared risk and returns (Sagalyn, 2007). According to Reynolds and Savage (2005), it became the norm for local communities to charter special redevelopment agencies, with and without eminent domain authority, to procure private property repurposed for economic development. (Schilling and Pinzón, 2016).

With that emphasis on leveraging private support, many post-1980 urban programs turned away from direct grants-in-aid, preferring to use tax credit programs that incentivize the private sector to invest in community assets. In many cases, those carrying out the projects, which may be not-for-profit organizations that don't get taxed, can't themselves take advantage of tax credits, so they sell the tax credits generated by the project to corporations or individuals who can use them. As one important example, the Low Income Housing Tax Credit (discussed in length in Chapter 4) was created as part of the Tax Reform Act of 1986. Through the LIHTC, about $8 billion a year in tax credits are leveraged to help finance affordable housing projects that mainly serve households earning no more than 60% of the area median income (the total amount of tax credits are limited and only a small percentage of eligible projects actually receive this support). This helps both nonprofit and for-profit developers increase the supply of affordable housing. This program has also proven resilient; guidelines have been amended by the program has survived many administrations. Critics note its complexity and the added costs incurred by the presence of a range of intermediaries (lawyers, accountants, syndicators) who are needed because of this complexity (https://www.taxpolicycenter.org/briefing-book/what-low-income-housing-tax-credit-and-how-does-it-work). But the complexity might also help account for the programs' political strength: Tax credit programs serve needy populations while also creating revenue for a number of professional services whose support no doubt keeps these programs funded. The Trump administration Opportunity Zone projects also function by providing tax benefits to the wealthy who invest in job-generating programs in designated areas.

Community Development in the 21st Century

The focus of federal community development policy since 2000 has remained centered on leveraging private resources, often through tax credit programs, in addition to the LIHTC described over, redevelopment experiments with new concepts such as New Market Tax Credits, Opportunity Zones, and synthetic TIF. They build on previous developer experiences and could serve to help future redevelopers (Cabanella, 2018). States in many cases have their own tax credit programs, most notably in area of brownfields remediation. Tax credits can help offset the costs of preparing a contaminated site for redevelopment and thereby make it easier to redevelopment core urban areas (Martin, 2018; Strom, 2018) (see Chapters 3 and 4 for further explanations).

But in addition to these private sector-focused programs, some federal programs from earlier eras continue to shape the community development policy arena. The public housing program, for example, may have its share of critics who point to often poor design and the tendency of public housing to exacerbate segregation by both race and income. According to HUD, some 970,000 low- to very low-income households live in public housing. Since the 1990s, many public housing authorities have become more entrepreneurial, bringing development partners in to help them redevelop aging housing projects into better designed, sometimes mixed income developments. The housing voucher program first authorized in 1937 and expanded in 1974 also continues to play a vital role in US communities, providing rent subsidies to low-income families. These can either be tied to a particular HUD project, or, more frequent, available to families to subsidize private market rentals. There are an estimated 5.21 million households with vouchers (https://www. cbpp.org/research/housing/national-and-state-housing-fact-sheets-data), clearly not enough to serve all qualifying low-income households.

In addition, the Community Development Block Grant continues to underpin the development of housing, community facilities, and related infrastructure in qualifying areas. In the most recent fiscal year, $3.4 billion was distributed among cities and states for community development activities (https://www.hud.gov/program_offices/comm_planning/budget). Because the CDBG distribution process is so well established and understood, the federal government has turned to that system to provide other kinds of aid to communities through that system. $5 billion of the 2020 CARES Act, passed to address emergency conditions arising from the spread of COVID-19, came to cities through a special addition to the CDBG program.

Conclusion

Throughout the 20th century and now two decades into the 21st century, the role of federal and local programs to combat urban decline and blight continues to be a key component of community development and urban policy. While early emphasis was placed on the heavy hand of the federal government to "fix" what ailed urban areas, there has now been a greater emphasis placed on the private sector through tax incentives and other subsidies.

It is the philosophical change in neo-liberal policies of both Republican and Democratic administrations that continue to see the private sector, and maybe more importantly the thinking of the private sector as can be seen in the shift of many housing authorities to developers in their own right, as being the ones to lead community development. It is this emphasis on the private sector mentality that is at the heart of current community real estate development thinking and in turn is the emphasis of much what is discussed in the preceding chapters.

Works Cited

81st Congress 1st, S. (1949). *Summary of Provisions of the National Housing Act of 1949*. Retrieved from Washington, DC: https://web.archive.org/web/20160215080101/, https://bulk.resource.org/gao.gov/81-171/00002FD7.pdf

Abrams, C., Lamb, T., & Robbins, I. S. (1936). New York City Housing Authority v. Muller. In Ny (Ed.), (Vol. 270, p. 333). New York County: Supreme Court.

Alker, S., Joy, V., Roberts, P., & Smith, N. (2000). The Definition of Brownfield. In (p. 49). Great Britain: CARFAX PUBLISHING CO.

American Federation of Housing Authorities, I. (1939). *State Housing Decisions: Summaries and Text*. Retrieved from Washington, DC: https://www.huduser.gov/portal/publications/state-housing-decisions-summaries-and-texts.html

Anderson, M. (1964). The Federal Bulldozer. A Critical Analysis of Urban Renewal, 1949–1962.

Arrington, G. B. (2018). Transportation. In B. Hersch (Ed.), *Urban Redevelopment: A North American Reader* (pp. 62–80). New York: Routledge.

Barrere, A., & Leland, C. G. (1889). *A Dictionary of Slang, Jargon & Cant*. London: Printed for subscribers only at the Ballantyne Press.

Bauer, C. (1934). *Modern Housing*. Boston, MA: Houghton Mifflin.

Bellush, J., & Hausknecht, M. (1967). *Urban Renewal: People, Politics, and Planning*. Michigan: Anchor Books.

Berman v. Parker, No. 22, 348 26 (Supreme Court 1954).

Bonastia, C. (2004). Hedging His Bets: Why Nixon Killed HUD's Desegregation Efforts. *Social Science History, 28*. doi:10.1215/01455532-28-1-19

Breger, G. E. (1967). The Concept and Causes of Urban Blight. *Land Economics, 43*(4), 369–376. doi:10.2307/3145542

Burgess, E. W. (2008). The Growth of the City: An Introduction to a Research Project. In J. M. Marzluff et al. (Eds.), *Urban Ecology* (pp. 71–78). Springer.

Cabanella, G. L. (2018). Revitalizing Neighborhoods, Housing and Social Equity. In B. Hersh (Ed.), *Urban Redevelopment: A North American Reader* (pp. 113–124). New York: Routledge.

Chicago, M. H. C. o. (1935). *Finding New Homes for Families Who Will Leave PWA Reconstruction Areas in Chicago*. Retrieved from Chicago, IL.

Clarke, S., & Gaile, G. (1999). Post-Fedezl Local Economic Development Policies. In L. R. J. Blair (Ed.), *Approaches to Economic Development: Reading from Economic Development Quarterly*. Thousand Oaks, CA: Sage.

Cook, J. F. (1959). The Battle against Blight. *Marquette Law Review, 43*, 444.

Ebenstein, W. (1938). The Law of Public Housing. *Marquette Law Review, 23*, 879.

Eisenhower, D. D. (1960). *Dwight D. Eisenhower: Containing the Public Messages, Speeches, and Statements of the President, 1953–61*. US Government Printing Office.

Farrar, M. E. (2008). *Building the Body Politic: Power and Urban Space in Washington, DC*: University of Illinois Press.

Fogelson, R. M. (2001). *Downtown: Its Rise and Fall, 1880–1950*: Yale University Press.

Fordham, J. B. (1949). Urban Redevelopment.

Fullilove, M. T. (2001). Root Shock: The Consequences of African American Dispossession. *Journal of Urban Health, 78*(1), 72–80.

Gans, H. J. (1982). *Urban Villagers*: Simon and Schuster.

Gordon, C. (2003). Blighting the Way: Urban Renewal, Economic Development, and the Elusive Definition of Blight. *Fordham Urban Law Journal, 305*. Retrieved from http://ezproxy.lib.usf.edu/login?url=http://search.ebscohost.com/login.aspx?direct=true&db=edshol&AN=edshol.hein.journals.frdurb31.24&site=eds-live

Government, U. S., Committee on, B., Currency, Senate, U. S., Subcommittee of Committee on, B., & Currency. (1931). *Creation of a Reconstruction Finance Corporation. [Electronic Resource]: Hearings before the United States Senate Committee on Banking and Currency, Seventy-Second Congress, first session, on Dec. 18, 19, 21, 22, 1931.* Retrieved from Washington, DC: http://ezproxy.lib.usf.edu/login?url=http://search. ebscohost.com/login.aspx?direct=true&db=cat00847a&AN=usflc.022249610& site=eds-live http://congressional.proquest.com/congcomp/getdoc?HEARING-ID =HRG-1931-BCS-0010

Green, T. L. (2018). Evaluating Predictors for Brownfield Redevelopment. *Land Use Policy, 73*, 299–319. doi:10.1016/j.landusepol.2018.01.008

Gries, J. M., & Ford, J. (1932). *President's Conference on Home Building and Home Ownership.*

Hersh, B. (2018). Chapter 1 History and Trends. In B. Hersh (Ed.), *Urban Redevelopment: A North American Reader* (First edition. ed., pp. 1–20). New York: Taylor & Francis.

Hill, P. H. (1952). Recent Slum Clearance and Urban Redevelopment Laws. In (Vol. 9, pp. 173–189).

Hill, R. C. (1983). Market, State, and Community: National Urban Policy in the 1980s. *Urban Affairs Quarterly, 19*(1), 5–20.

Hipler, H. M. (2007). Economic Redevelopment of Small-City Downtowns: Options and Consideration for the Practitioner. *The Florida Bar Journal, 81*, 39.

Hurd v. Hodge, No. 68 S.Ct. 847 (Supreme Court of the United States 1948).

Jacobs, J. (1961). *The Death and Life of Great American Cities* (First ed.). New York: Vintage Books.

Johnstone, Q. (1958). The Federal Urban Renewal Program. *The University of Chicago Law Review, 25*(2), 301–354. Retrieved from https://www.jstor. org/stable/1598136?seq=1

Kelo v. New London, No. 04-108, 545 469 (Supreme Court 2005).

Leigh, N. G. (2003). *The State Role in Urban Land Redevelopment*: Brookings Institution Center on Urban and Metropolitan Policy.

Logan, J. R. (1983). The Disappearance of Communities from National Urban Policy. *Urban Affairs Quarterly, 19*(1), 75–90.

Maltbie, T. M. (1944). Urban Redevelopment. *Yale Law Journal, 54*(1), 116–140. Retrievedfromhttps://digitalcommons-law-yale-edu.ezproxy.lib.usf.edu/ylj/vol54/ iss1/14

Martin, G. D. (2018). *The Aftermath of Redevelopment Agencies: A Case Study on the Abolishment of Redevelopment Agencies and the Impact It Has Had on Economic Development in the County of Riverside, California*, University of La Verne, (Dissertation/Thesis).

Massachusetts v. Mellon, No. 24. Original, 262 447 (Supreme Court 1923).

McKenzie, R. D. (1924). The Ecological Approach to the Study of the Human Community. *American Journal of Sociology, 30*(3), 287–301.

Milgram, G., Library of Congress Congressional, R. S., United States Congress House Committee on Banking, F., Urban, A., United States Congress House Committee on Banking, F., Urban Affairs Subcommittee on, H., & Community, D. (1994). *A chronology of housing legislation and selected executive actions, 1892–1992: Congressional Research Service report prepared for the Committee on Banking, Finance and Urban Affairs and the Subcommittee on Housing and Community*

Development, House of Representatives, One Hundred Third Congress, first session: U.S. G.P.O.

Moore, S. W. (2005). Blight as a Means of Justifying Condemnation for Economic Redevelopment in Florida. *Stetson Law Review*, 443. Retrieved from http://ezproxy.lib.usf.edu/login?url=http://search.ebscohost.com/login.aspx?direct=true&db=edshol&AN=edshol.hein.journals.stet35.24&site=eds-live

Morris, E. J. (1974). The Nixon Housing Program. In (Vol. 9, p. 2): *Section of Real Property, Probate and Trust Law*. American Bar Association.

National Conference on City, P. (1924). Proceedings of the ... National Conference on City Planning. In Boston: The Conference.

Nesbitt, G. B. (1949). Relocating Negroes from Urban Slum Clearance Sites. *Land Economics, 25*(3), 275–288.

Nuissl, H., & Heinrichs, D. (2013). Slums: Perspectives on the Definition, the Appraisal and the Management of an Urban Phenomenon. *DIE ERDE–Journal of the Geographical Society of Berlin, 144*(2), 105–116.

Pritchett, W. E. (2003). The Public Menace of Blight: Urban Renewal and the Private Uses of Eminent Domain. In (Vol. 21, pp. 1–52).

Rast, J. (2012). Why History (Still) Matters: Time and Temporality in Urban Political Analysis. *Urban Affairs Review, 48*(1), 3–36.

Records, F. F. F. C., & New York Public Library, M. a. A. D. (1920s). Backyard Privy, Sole Toilet Facilities for and Apartment House. In New York.

Reingold, D. A. (2001). Are TIFs Being Misused to Alter Patterns of Residential Segregation? The Case of Addison and Chicago, Illinois. *Tax Increment Financing and Economic Development: Uses, structures and impact, State University of New York Press, Albany*, 223–241.

Reissman, L. (1964). The Urban Process: Cities in Industrial Societies.

Reynolds, R., & Savage, D. G. (2005). Property Ruling Strikes Nerve in House. *Los Angeles Times*. Retrieved from https://www.latimes.com/archives/la-xpm-2005-jul-01-na-property1-story.html

Richter, I. (1983). *The Evolution of Federal Housing Programs and Their Impact on the District of Columbia*: Department of Urban Studies, University of the District of Columbia.

Roberts, P., (2016). The Evolution, Definition and Purpose of Urban Regeneration. In P. Roberts et al. (Eds.), *Urban Regeneration: A Handbook* (pp. 9–43). Thousand Oaks CA: SAGE Publications.

Rothstein, R. (2017). *The Color of Law: A Forgotten History of How Our Government Segregated America* (First edition. ed.). Liveright Publishing Corporation, a division of W.W. Norton & Company.

Sagalyn, L. B. (2007). Public/private Development: Lessons from History, Research, and Practice. *Journal of the American Planning Association, 73*(1), 7–22.

Schacht, W. (2018). Urban Design and City Form in Redevelopment. In B. Hersh (Ed.), *Urban Redevelopment: A North American Reader* (pp. 42–61). New York: Routledge.

Schilling, J., & Pinzón, J. (2016). The Basics of Blight. Recent Research on Its Drivers, Impacts, and Interventions. *VPRN Research & Policy Brief* (2).

Senate, U. S. (1937). United States Housing Act of 1937. 75th Cong., 1st sess. *S, 1685* (1974), 3006.

Shelley v. Kraemer, No. 334 U.S. 1 (Supreme Court of the United States 1948).

Social Welfare History, P. (2018). [Tenement house reform. Social Welfare History Project]. Web Page.

States, U. (1953). *Recommendations on Government Housing Policies and Programs, a Report*. Washington: USHUD.

Stoloff, J. A. (2004). *A Brief History of Public Housing*. Paper presented at the annual meeting of the American Sociological Association, San Francisco, CA.

Strom, E. (2018). Brownfield Redevelopment: Recycling the Urban Environment. In R. Brinkmann & S. J. Garren (Eds.), *The Palgrave Handbook of Sustainability: Case Studies and Practical Solutions* (pp. 371–384). Cham: Palgrave Macmillan.

Sutton, S. A. (2008). Urban Revitalization in the United States: Policies and Practices. *United States Urban Revitalization Research Project (USURRP)*.

Teaford, J. C. (2000). Urban Renewal and Its Aftermath. In (p. 443). United States: HOUSING RESEARCH.

Tepper, R. B. (2001). A Thousand Points of Blight. *Los Angeles Lawyer, 34*.

The City of New, Y. (1901). *The Tenement House Law of the City of New York, with Headings, Paragraphs, Marginal Notes and Full Indexes. The Record and Guide*. New York: The Record and Guide.

Thompson, W. (1903). *Housing Handbook*. London: National Housing reform Council.

TIME. (1934). Knickerbocker Village, *24*(Generic), 18. Retrieved from http://ezproxy.lib.usf.edu/login?url=http://search.ebscohost.com/login.aspx?direct=true&db=edb&AN=54804275&site=eds-live

United Nations Human Settlements Programme, S. (2003). *The Challenge of Slums: Global Report on Human Settlements, 2003*. UN-HABITAT.

United States. Bureau of, L., Houghton, A. S., & Olmsted, V. H. (1894). The Slums of Baltimore, Chicago, New York, and Philadelphia: prepared in compliance with a joint resolution of the Congress of the United States. In: Govt. print. off.

United States. President's Homes, C., & Sternberg, G. M. (1909). Reports of the President's Homes Commission. In: U. S. Govt. Print. Off.

United States: National Housing Agency, O. o. t. G. C. (1947). *Comparative Digest of the Principal Provisions of State Urban Redevelopment Legislation*. Retrieved from Washington D.C.: https://www.google.com/books/edition/_/xak-UGpy34AC?hl=en

Vaux, J. H., & Field, B. (1819). *Memoirs of James Hardy Vaux*: W. Clowes.

von Hoffman, A. (2000). A Study in Contradictions: The Origins and Legacy of the Housing Act of 1949. *Housing Policy Debate, 11*(2), 299–326. doi:10.1080/10511482.2000.9521370

Walker, M. L., & Wright, H. (1938). *Urban Blight and Slums: Economic and Legal Factors in their Origin, Reclamation, and Prevention* (Vol. 12): Harvard University Press.

Weber, R., & O'Neill-Kohl, S. (2013). The Historical Roots of Tax Increment Financing, or How Real Estate Consultants Kept Urban Renewal Alive. *Economic Development Quarterly, 27*(3), 193–207.

Webster, I. (2020). [CPI Inflation Calculator]. Web Page.

Winslow, C.-E. (1937). Housing as a Public Health Problem. *American Journal of Public Health and the Nations Health, 27*(1), 56–61.

Wood, E. E. (1969). *Slums and Blighted Areas in the United States*. College Park, Md.: McGrath Pub. Co.

Wortzel, A. (1997). Greening the Inner Cities: Can Federal Tax Incentives Solve the Brownfields Problem 13th Annual R. Marlin Smith Student Writing Competition Award Winner. *Urban Lawyer, 29*(2), 309–340. Retrieved from https://heinonline.org/HOL/P?h=hein.journals/urban29&i=332, https://heinonline.org/HOL/PrintRequest?handle=hein.journals/urban29&collection=journals&div=26&id=332&print=section&sction=26

Zhang, Y., & Fang, K. (2004). Is History Repeating Itself? From Urban Renewal in the United States to Inner-City Redevelopment in China. *Journal of Planning Education and Research, 23*(3), 286.

3 Programs to Make the Numbers Work and Their Impacts

Low-Income Housing Tax Credits, New Market Tax Credits, and Opportunity Zones

Stephen T. Buckman

Introduction

Real estate development on the surface is about remaking the landscape and creating buildings and areas that help urban environments thrive. Yet, while real estate development is about remaking the urban landscape, it is also of course about making money. Real estate developers have their names on buildings and even become president of the United States for a reason. Developers by their nature are not benevolent beings; while many may try to do the right thing, they still need to make a development financially viable. Real estate development like any other business is a business that needs to make a profit and while society increasingly is demanding more affordable housing, developers still have equity partners, banks, GCs (general contractors), and sub-contractors down the line that need to be paid. While many developers may want to make their developments more affordable or community-driven, their financial obligations and guarantees may say otherwise.

How then do developers who wish to do more community and affordable housing and yet need to make a required return get beyond the crossroads that they sit at? For developers that want to be able to do both, they must be blessed with one of two things. One they would need either to be independently wealthy or have a wealthy benefactor that believes in societal good over robust returns aka a benevolent "thanks Dad." Or two they need to have outside help from governments to mitigate the financial hardship that comes with foregoing 10–20% or more of their market rate development to community or affordable housing. While ideally, we all would be fortunate to find ourselves in the former most often than not, we find ourselves in latter.

Often the outside help that developers doing community and affordable housing type of work garner are subsidies from federal and local governments. The subsidies come in many forms from streamlined permitting to density bonuses to direct financial payment. Yet, the most often and most widely sought after and used are tax credits and/or tax sheltering subsidies. On the credits side, such as Low-income Housing Tax Credits (LIHTC) and New Markets Tax Credits (NMTC), these will come in the form of the

DOI: 10.1201/9781003109679-4

allocation of tax credits given to the developer and then sold on the open market at a rate such as 85% of a dollar or $0.85 for each dollar awarded, depending on the market at the time of sale, which then allows the developer to use that money as equity. On the other hand, developers can look to develop in designated zones, such as an Opportunity Zone (OZ), which allows tax burdened investors the opportunity to invest in a development and defer their tax penalty for a number of years in turn allowing developers to use that money as equity. Another important tax credit which will not be covered here but is covered by Bruce Steifel in Chapter 6 is Historical Tax Credits which give developers tax credits to sell in order to do redevelopment of historical properties.

Subsidies such as these then become the lifeblood for affordable and community real estate development developers. Without these subsidies, these developers would not be able to get the numbers to work to allow a development to move forward. In developer lingo, the "project would not pencil out." And as governments are demanding more community and affordable housing, it is in their best interest to continue to help maintain and spur more subsidies of this nature.

Understanding the need and importance of subsidies in the affordable housing market, this chapter will highlight three key forms that have been commonly used and what has the potential to be the most widely used. In this chapter, I will first discuss the need for developer subsidies. I will follow that with an overview of LIHTC, NMTC, and OZs and how they are used in a real estate development deal.

Subsidies: Why We Need Them

The driving force behind much of what CRED stands for is the issue of equity and to establish equity, it is important people have access to affordable housing and good neighborhoods. Thus, there is a need for more affordable housing to begin with. But the issue at hand is that there is a substantial lack of affordable and workforce housing. While many developers may want to include affordable and workforce housing in their developments without the previously mentioned subsidies, this is not a reality.

Affordability is a huge issue in the United States. According to the National Low Income Housing Coalition (NLIHC), there is a shortage of roughly seven million affordable homes for the nation's 11 million low-income families, noting that 75% of all extremely low-income families are severely cost-burdened which is defined as paying more than half their income on rent (Gramlich, 2021a, 2021b, 2021c) while a basic level of being cost-burdened is defined as more than 30% (Habitat for Humanity, 2021). Further exacerbating this issue of income as a percentage of rent is that it is widening each year. For instance, today's real average wage – after accounting for inflation – has about the same purchasing power as it did 40 years ago, with home prices rising faster than inflation over the last 25 years: home prices up 41% and inflation up 20% (Habitat for Humanity, 2021).

The housing challenge became even more extreme during COVID-19 which resulted in a rise in working from home. This emphasis on white-collar workers being able to work from home, note this was not an option for most blue-collar and minimum wage workers or those most in need for affordable and workforce housing, resulted in an increase in housing costs in many communities that had never experienced these shocks before. A case in point is that from June 2020 to June 2021, we saw a jump from the median US home price being $294,000–$363,000 which equates to a year-to-year jump of 23.4% (Lane, 2021).

The results of this radical jump of 23.4% and higher housing prices were caused mainly due to two dynamics. The first being that as people worked from home, they decided that they needed/wanted more space. Secondly, many people who were forced to live in certain locations to be close to places for work were no longer wedded to the anchor of where their work was, and thus, they could almost move anywhere. This was further stoked by the federal government with ultra-low and artificially reduced interest rates and the general COVID-19 Federal Stimulus package (Lane, 2021).

To put the need for affordable housing into perspective, Greenville, SC poses as a perfect case study. Greenville, which was historically a blue-collar textile town that fell on bad times, has over the last 20 years reframed itself as one of the top small Southern downtowns and in turn has had massive influx of development and money to the area. With this influx of new money, there has been rapid gentrification, especially in downtown which has resulted in rents and housing prices that are now only affordable for a few. Greenville had been a town where one could afford to live on a modest salary and buy a modest home, and those days are over in Greenville.

For instance, in Greenville, there are 9,546 households – roughly 30% of the city's 32,250 total households – that make below 50% AMI (Contino, 2021). This results in huge housing disparities for those that are not of privilege as the average rent for a one bedroom in Greenville is $770 which means someone who is working a minimum wage job would have to 79 hours a week to not be cost-burdened. Thus, Greenville needs to build 10,000 or more units of affordable housing, meaning homes for those that are at 80% AMI to combat the problem (Contino, 2021). The only way that this many homes could come close to being built and in turn neighborhoods stabilized is through government incentives such as LIHTC, NMTC, and OZs.

Low-Income Housing Tax Credit (LIHTC)

Low Income Housing Tax Credits or better known by the acronym of LIHTC is one of the most widely used tax incentives that developers use to offset the costs of foregoing market rate units in the name of affordable housing. LIHTC allows developers to apply for and obtain credits that they can sell at a percentage in the open market. The proceeds of the sale for the developer equals equity and for the investor, it equals the ability to reduce their tax liability.

The program started as part of 1986 Tax Reform Act and looks to help developers construct developments for low-income renters. The program can be used for various different projects. These projects include: "multifamily or single-family housing; new construction or rehabilitation; special needs housing for elderly people or people with disabilities; and permanent supportive housing for homeless families and individuals" (Hoffman, 2017: 5–30).

How it works: The state or municipality applies and is awarded a certain number of tax credits by the IRS. These credits are awarded on a need-based basis in that the municipality needs to show that the citizens are lower income and in need of subsidization. Once the state or municipality has been awarded credits, developers then can apply for those credits based on certain affordable housing stipulations (more discussion on stipulations to follow). Assuming the developer has checked all the required boxes and is eligible the developer is then awarded a certain number of credits depending on various impact metrics such as size, need of community, amount of affordable housing units, etc.

The developer will take their awarded credits, for example, say they were awarded $1 mm in tax credits, and look to sell them on tax credit market. The credits then will be sold at a percentage of their full cost to a syndicate, company, or high net worth individual that is looking to reduce their tax burden. That syndicate will buy the credits at a percentage, for example, say $0.85 on the dollar, thus paying $850,000 for the credits. In this scenario, the developer will get $850,000 in money for the credits which they will then use for equity and the syndicate (those with the tax burden) will lessen their tax burden by $150,000. Breaking this down further in this scenario, the syndicate will get $100,000 year of tax credits for ten years ($1 mm in total spread over ten years equally) and the developer will get the equity which they then will keep their building LIHTCed or a number of units affordable for 15 years.

For the developer to obtain these credits, they must include estimates of the expected costs of the project and commitment to adhere to one of two conditions. Option one at least 20% or more of the residential units must be rent-restricted and occupied by individuals whose income is 50% or less than the area median gross income. The second option the developer could pursue is that at least 40% or more of the residential units in the development are both rent-restricted and occupied by individuals whose income is 60% or less of the area median gross income. Note developers are not restricted to 40% or 20% of their units and there is nothing stopping a developer from having 100% of the units be rent-restricted, which happens often. Also, it is important to note that there are no limits on the amount the developer can charge for units that are not rent-restricted.

There are basically two types of LIHTC credits available to the developer: 9% or 4% credits. The 9% credit can cover 70% of the development's equity, while the 4% can cover up to 30% of the development's equity; this is often referred to as the 70/30 rule. The difference between the two types, in its simplest description, is that the 9% credit is usually for new construction or rehabilitation without any other federal subsidies which is a key component in a key stipulation. While the 4% is for the acquisition of existing buildings, rehabilitation, and/or new construction but unlike the 9%, the 4% is eligible to obtain other federal subsidies such as Historical Tax Credits, NMTCs, etc. More often developers will obtain the 4% credit as the stipulations are less stringent and it leaves them the option to obtain other types of subsidies.

Issues with LIHTC

The main problem with these incentives that are originally made to benefit the low-income residents is lack of guaranteed benefits for low-level income communities in poor areas and lack of community-oriented tax benefits under the current law (Layser, 2019). Moreover, tax credit incentives are administered federally and mostly include individual buildings rather than an entire neighborhood, which benefits private developers more than communities (McCabe, 2020). A property tax credit may also stimulate housing supply and cause changes in amenities, such as drawing more businesses, which eventually affect the rental prices (Baum-Snow and Marion, 2009). Field (2016) declared that investors are more likely to invest in gentrifying neighborhoods as the probability of getting tax credits is higher. However, incentives tax credit can more meaningfully impact poor and non-gentrifying neighborhoods and are a better way of using federal resources.

The issue of gentrification is a substantial one as it pertains to LIHTC. Gentrification which often is a byproduct, or sometimes a direct intention of the real estate industry (see Chapter 9) is further stoked with the use of LIHTC. While the use of LIHTC is highly beneficial to both the developer and to low-income individuals in the front end of the development, it has the contrary impact in the backend when the IHTC credits roll off, in 15 years. While it can be beneficial for the developer who will now be able to sell a stabilized building in a most likely re-established neighborhood, for the low-income renter, it will most likely mean eviction as the new owner of the building, more times than the present owner will turn the building into market rate when the credits roll off in turn evicting the lower income tenant who cannot afford to pay market rate. Adding to the situation is that the neighborhood that person had called home is unaffordable as well as it has become gentrified.

LIHTC Example

An example of how LIHTC could play out is as follows. There is a 70 unit development of which 40% are income and rent restricted. In regard to the project, there are no other federal grants used to help finance the project.

Table 3.1 LIHTC Development Example

Total Development Costs	$5,000,000
Land acquisition	$1,000,000
Construction	$3,400,000
Site improvement	$535,000
Engineering	$40,000
Eligible soft costs	$25,000

Eligible Basis: Total Development Cost − Land Acquisition= $4,000,000
Qualified Basis: Eligible Basis × Applicable Fraction ($4,000,000 × 0.40) = $1,600,000
Annual Tax Credit: Qualified Basis × Tax Credit Rate ($1,600,000 × 0.09) = $144,000
Total Amount for Tax Credits: $144,000 × 10 years = $1,440,000

The developer has received the tax credits and has put them out to the market to sell. The developer has been able to obtain $0.75 for each tax credit (the tax credit price). Using this scenario, the developer would get $1,080,000 ($1,440,000 × 0.75) equity for a ten year stream of future benefits amounting to $1,440,000 ($144,000 a year) for the syndicate (see Table 3.1).

New Market Tax Credits (NMTC)

While LIHTC allows developers to offset the losses from non-market residential multi-family units, it also becomes important to encourage and allow for retail development in underserved and in turn perceived less profitable areas. To encourage retail development and to offset the perceived lesser returns of developing in low-income communities, like residential development, it becomes necessary for developers to find a way to bridge the financial return gap. Much like LIHTC, NMTC allows developers to obtain tax credits and then sell those credits on the open market.

The program, following in the path of programs such as enterprise zones and economic OZs, was enacted in December 2000 as a part of the Community Renewal Tax Relief Act of 2000 to revitalize low-income communities. Much like LIHTC, NMTCs are federal tax credits to assist in the funding of neighborhood changing/job creating commercial real estate projects and businesses located in low-income census tracts. The driving force for this program is the notion that low-income communities have difficulty attracting investment, leading to dormant or vacant buildings and businesses, inadequate access to healthcare and education, lower property values, contributing to low-income communities becoming 'distressed' communities. Thus, NMTCs are there to encourage developers to look to these communities.

As stated, a key component is the notion of a Low-income community (LIC). These LICs are determined by census tracts within those census tracts income levels must be accomplished. These levels include minimum or higher rating with the awarding being based on the emphasis on severity and need. Minimum qualifications for the census tract include 80% or lower median family income, 20% or higher poverty and a targeted population. While the minimum qualifications is just that the minimum there are two ways determining higher levels of need: (1) Severe (one of the following) 30% poverty; 60% median family income; 1.5× unemployment rate; or non-metro; (2) Higher distressed (two of the following): 25% poverty, 70% median family income, or 1.25x unemployment rate and in an SBA Hub Zones, brownfield, HOPE VI area, Federal Native area, Federally Medically Underserved area, State/Local Economic Zones, FEMA Disaster area or a food desert.

Once it has been determined that your project falls into one of these characteristics, it becomes important that the retail establishment you want to have in your development is allowable by NMTC standards. So, for your project to qualify, it must first be located in LIC census tract, and secondly, it cannot be any of the following: residential rental property, buildings that derive 80% or more of income from residential dwelling units, and Certain Business Type (including: business consisting predominately of developing/holding intangibles, golf courses/country clubs, certain farming businesses, massage parlors, hot tub facility, tanning facility, gambling facilities, and ABC stores/alcohol stores). It will be important that you work closely with your NMTC consultant to make sure not only you fit in one of the LICs but also the businesses you seek to fill your development adhere to these rules.

So then what constitutes a good candidate. First is that your development is located in a "severely distressed" census tract (see stipulations mentioned earlier). The second is that Compelling community impact means that there is a tangible community benefit – measured by quality job creation, providing needed goods and services to low-income communities (grocery stores, medical facilities, charter schools, etc.), as part of an existing plan for economic revitalization and/or that NMTC fills a real funding gap that would prevent project from moving forward also known as the "but for test." The third characteristic that makes you a strong candidate is that it is "ready to go," meaning other sources of funding are committed to the project and the necessary approvals are all in place.

Issues with NMTC

Based on the initial purpose, the program intends to benefit the low-income residents other than benefiting the wealthy investors who come into low-income neighborhoods. Thus, the main issue is that the investors are not necessarily local residents. The program aims to enhance the economic development of a geographic area other than the individuals living in the area and therefore, it benefits people who came from outside of the area more (Groves, 2019). Groves (2019) introduces the "pure people-oriented" and "people-in-place oriented" strategies. The former aids people regardless of where they live. It targets the geographic area, regardless of the relationship to the LIC. This is what is happening with New Market Tax Credit, which Groves (2019) believed that it is ill-conceived as a facilitator to mainly benefit low-income households within the poor areas, as Congress intended (Groves, 2019). The remedy, however, is redirecting funds to more precisely benefit existing low-income residents who are the main object of the program (Groves, 2019).

Therefore, Groves (2019) argued that because of the policies of this program, the consequence is marginalization or squeezing the existing residents out of the neighborhoods. Some examples of long-term effects of this program are Portland Oregon, which lost the black community, and Seattle, where the black community were replaced with well-educated and mostly white newcomers (Groves, 2019).

Example Case

As an example, we can look at a possible $10mm project. In this scenario, the developer is building a new facility and purchasing new equipment. The project meets all the NMTCs requirements, LIC, and strong community impacts, in turn making it eligible for NMTC financing. As mentioned, the project highlights the "there but for" clause in that the project would not be able to get done but for the credits. In this scenario, the project would receive $1.5 mm in sponsor capital i.e. equity and $5,185,000 in loans/grants. This results in project sources of $6,685,000 leaving them with a shortfall of $3,315,000 as the project uses or costs $10mm. To close this shortfall, the developer would apply for NMTC via a CDE (Community Development Entity). The developer is able to receive $3.9mm in credits which the CDE was able to broker the sale of the credits at $0.85 to the dollar resulting in $3,315,000 (see Table 3.2).

It is important to note unlike LIHTC, NMTC are handled through a CDE. The CDE is allotted so much in NMTCs for their organization and will look to get as much as they can for the credits on the open market. For instance, if the CDE was able to get $0.95 to the dollar, they would have only had to sell $3,489,474 of their allotted credits leaving them with $410,000 to invest in another project. LIHTC, on the other hand, is awarded directly to the developer and thus if the developer is able to get a better price, they would keep the difference.

Table 3.2 NMTC Case Study: $10MM Project

Project Sources		Project Uses	
Sponsor affiliate capital	$1,500,000	Site acquisition/hard construction costs	$5,200,000
Loans or Grants	$5,185,000	Soft construction costs	$800,000
Total:	$6,685,000	Machinery and equipment	$3,000,000
		Financing costs	$1,000,000
GAP:	$3,315,000	Total	$10,000,000

Source: Whitney Ferguson Lecture (March 2019).

Opportunity Zones

The sexiest incentive program to come to the forefront of the community driven real estate in many years is the OZ program. At the time of this writing, it is still in its infancy, yet at the same time is garnering much attention and financial infusion. The Opportunity Zone program (OZ) has the chance to change the playing field and presents a very potent option for community real estate developers.

OZs came out of the Tax Cut and Jobs Act of 2017, as basically a rider to the larger bill with the intention to revive investment in communities that have experienced extensive poverty and distress. The OZ program is built on the backs of a long tradition of economic development policies such enterprise zones and economic development zones that look to lift communities up via the private sector and not public sector. The OZ was devised and the brainchild of the Economic Innovation Group with the intention to direct capital investment to areas that had not and have not recovered from the 2008 recession.

The intent of the program is similar to other programs, which is providing tax incentives for investments in the designated zones (Carter, 2019). They include low-income communities with the poverty rate of at least 20% or a median income less than 80% of the area median income (AMI) in their metropolitan area (Jacoby, 2019). Since its creation, 8,700 OZs in states and territories have been designated. The OZ provision provides investors with tax benefits in form of capital gains in the value of investments, such as stock, that have not been taxed yet by the government (Jacoby, 2019).

Investors can benefit from tax break for OZs in up to three ways:

> First, investors can defer taxes on their capital gains until 2027 if they invest their gains in opportunity zone funds. Second, those who hold their opportunity zone investments for at least seven years also will get a 15% cut in the capital gains taxes that they would otherwise pay (on top of the generous tax break that the low capital gains tax rate already gives them). Third, and perhaps most significant, those who hold opportunity zone investments for at least ten years will get a permanent capital gains tax exemption for all the gains they realize on their opportunity zone investments through 2047.
>
> (Jacoby, 2019: 2)

The investments in OZs must invest 90% of the raised capital in physical assets – such as real estate or equipment located in the OZs – and/or ownership interests – such as stocks that operate at least partially in OZs. The other 10%, however, is not subject to any restrictions (Jacoby, 2019).

The program works through the creation of an opportunity fund where investors place their capital gains taxes into a fund within 180 days. The fund then, as mentioned, must hold 90% of its assets for a fixed period of time:

- 5 years the investor earns 10% step-up basis;
- 7 years a 15%, step-up;
- 10 years or longer, the investor is qualified for an increase in the basis of the investment equal to its fair market value on that date the investment is sold (IRS Opportunity Zone Fact sheet, 2018).

These step-up ladders show that there is incentive for the investor to stay in the project for a longer period of time. For the developer, this means that the initial money they received through the fund to help fund their project will remain with the project and then the developer will not have to seek more funding or refinance the project. Furthermore, these monies allow the developer to offset the loss in review revenue that is not obtained by building in a transitional area (Table 3.3).

Issues with OZ

The issues with OZ program are the lack of community engagement, lack of Opportunity Fund certification by the US Treasury, high risk of displacement, lack of protective policies, the competition between local Community Development Financial Institutions and larger national financial institutions without local knowledge, lack of tangible recording, the flexibility and lack of finalized regulations (Carter, 2019), diverting investment from truly disadvantaged communities, mainly benefiting wealthy investors instead of residents of these areas, reducing federal revenues, and creating new opportunities for tax avoidance (Jacoby, 2019). Moreover, Jacoby (2019) argued that because of the rapid drafting and passing of the 2017 law, without public participation or expert input, the program did not receive the attention it deserved. He also criticized the regulation and noted that:

Table 3.3 The Incentive Structure

Incentive Structure

The OZ program offers three tax incentives for investing in low-income communities through a QOF:
1. **Temporary deferral.** The deferral of inclusion in taxable income for capital gains reinvested in a QOF. The deferred gain must be recognized on the earlier date on which the opportunity zone investment is disposed of or on December 31, 2026.
2. **Step-up in basis.** Capital gains reinvested in a QOF. The basis is increased by 10% if the investment in the QOF is held by the taxpayer for at least 5 years and by an additional 5% if held for at least 7 years, thereby excluding up to 15% of the original gain from taxation.
3. **Permanent exclusion.** The exclusion from taxable income of capital gains from the sale or exchange of an investment in a QOF if the investment is held for at least 10 years. This exclusion only applies to gains accrued after an investment in a QOF.

Source: Ortiz (2021).

This definition of "low-income community" is broad enough to include some areas that are not truly distressed, such as areas adjacent to some elite colleges — for example, the University of Virginia and the University of California at Berkeley, where a large concentration of students skews the income data. As noted, the law requires that opportunity zone businesses "derive" at least half of their income from an "active business" in a zone. Neither the law nor the regulations, however, explain what that means. Opportunity zones also bring the potential for loopholes that encourage tax sheltering and other forms of tax avoidance. As a starting point, the proposed regulations let an investor get the full tax break even if only 63 percent of the total capital that an opportunity zone fund invests flows to a zone. The direct tax benefits of opportunity zones will flow overwhelmingly to wealthy investors, but the tax break might not do much to help low-income communities, and it could even harm some current residents of such communities.

(Jacoby, 2019: 5–6)

However, Jacoby (2019) noted that even though the program assures investors about accumulating more wealth, there are no requirements to ensure that low-income residents of designated areas benefit from investments. Therefore, the OZ tax break may lead to a "subsidy for gentrification instead of providing housings and jobs for low-income residents of these areas."

To make the program work better, researchers suggested some recommendations such as more restrictive regulations to reduce the possibility of the program serve as pure tax shelter, less ambiguity in what makes qualified activities in an OZ (Jacoby, 2019), better understanding of the community members and local business owners, more transparency in the process, maintaining public domains, and protecting existing new construction of affordable housing (Carter, 2019).

Example

In further research from Eastman and Kaeding (2021), Table 3.4 (A) below illustrates the tax benefits that an OZ will bring to an investor who invests in a QOF for 7 years. This hypothetical investor made a $1.5 million profit after investing $1 million and selling it for $2.5 million. They will owe $357,000 in capital gains taxes under the old scheme. The investor would save $53,550 in capital gains taxes if they invested the $2.5 million in a QOF. Because of the step-up in basis, they now have to pay taxes on $1.275 million in capital gains rather than the initial $1.5 million. Table 3.4 (B) now shows that any benefit realized after the $1.5 million investment in the QOF is tax-free if kept for at least 10 years. The investor would save an extra $119,000 in taxes if he or she made another $500,000 over the 10 years of the QOF investment. By reinvesting their $1.5 million capital gain in an OZ and retaining it for 10 years, this hypothetical investor saves $172,550 overall in taxes.

Table 3.4 Example Tax Benefit Deal

A

Sample Calculation of Capital Gains Tax on a QOF Investment Held for seven years

Traditional Investment		QOF Investment Held for seven years	
Original investment:	$1,000,000	Original investment:	$1,000,000
Sold for:	$2,500,000	Sold for:	$2,500,000
Capital Gain:	$1,500,000	Capital Gain Transferred to QOF:	$1,500,000
		Increased Basis:	$1,225,000
		New Calculated Gain:	$1,275,000
Capital Gain Tax Rate:	23.80%	Capital Gain Tax Rate:	23.80%
Capital Gain Tax Due:	$357,000	Capital Gain Tax Due:	$303,450
Post-Tax Earnings:	$2,143,000	Post-Tax Earnings:	$2,196,550
		Tax Savings:	$53,550

Note: Investors can increase their original cost basis, $1M, by 15% of their original gain that was transferred ($1.5M) into the QOF if the investment is held for seven years. This calculation does not include any earnings made from the QOF investment.

Source: Taxfoundation.org

B

Sample Calculation of Capital Gains Tax on Accrued Earnings of QOF Investment Held for ten years

Traditional Investment		Earnings Accrued from QOF Investment	
Original investment:	$1,500,000	Original investment:	$1,500,000
Sold for:	$2,000,000	Sold for:	$2,000,000
Capital Gain:	$500,000	Capital Gain Transferred to QOF:	$500,000
Capital Gain Tax Rate:	23.80%	Capital Gain Tax Rate:	0.00%
Capital Gain Tax Due:	$119,000	Capital Gain Tax Due:	$
Post-Tax Earnings:	$381,000	Post-Tax Earnings:	$500,000
		Tax Savings:	$119,000

Note: Any gain accrued after the $1.5 million investment in the QOF is exempt from taxation if it is held for ten years or longer. Assuming the investor made another $500,000 over the decade of the QOF investment, the investor would save an additional $119,000 in taxes.

Source: Taxfoundation.org

Source: Taxfoundation.org in Ortiz (2021).

Conclusion

As discussed in this chapter, credits such as LIHTC and NMTC and tax shelters such as OZs are the lifeblood of any affordable housing or CRED developer. It is one of the main ways that a developer can offset the loss they take for engaging in CRED oriented development. Yet, they may be able to make the proforma numbers pencil out by engaging in these subsidies resulting in benefits in the front end of the development for themselves and yet oftentimes for their clients/residents' trouble results in the back end of the development. This trouble is often manifested in issues of gentrification, ironically creating more issues of inequity which was what they were originally looking to offset.

While subsidies of this nature are important to offset development costs making CRED projects doable, they also come with drawbacks beyond issues such as gentrification. The biggest and most unsettling to the developer is the fact that anytime money is taken from the government, the developer's dealing must become more transparent which many developers battle with. Not implying they are doing something nefarious but rather the red tape they must cut through and keeping certain financial things close to the vest in relation to completion. In the end, it may not be worth it.

Works Cited

Baum-Snow, N. and Marion, J. (2009) "The Effects of Low Income Housing Tax Credit Developments on Neighborhoods" In *Journal of Public Economics*. Vol. 93(5–6): 654–666.

Carter, M. (2019) *Federal Opportunity Zones: The Newest Gentrification Tool?* Georgia Tech School of City and Regional Planning Applied Research Paper.

Contino, G. (6.30.2021) "The 'Affordable Housing' Planned in Greenville Largely Ignores Those Who Need It Most" In *Greenville News*.

Eastman, S. and Kaeding, N. (9.04.2021) *Opportunity Zones: What We Know and What We Don't*. Tax Foundation. https://taxfoundation.org/opportunity-zones-what-we-know-and-whatwe-dont/

Fikri, K. and Lettieri, J. (12.2018) *The State of Socioeconomic Need and Community Change in Opportunity Zones*. Economic Innovation Group.

Gramlich, E. (2017a) "Public Housing" In *Advocate's Guide*, National Low Income Housing Coalition: Washington, DC.

———. (2017b) "Housing Choice Vouchers" In *Advocate's Guide*, National Low Income Housing Coalition: Washington, DC.

———. (2017c) "Low Income Housing Tax Credits" In *Advocate's Guide*, National Low Income Housing Coalition: Washington, DC.

Habitat for Humanity. (2021) *What Is Housing Affordability?: Cost of Home*. www.habitat.org/costofhome/what-is-housing-affordibility. Retrieved: 8.27.21.

Hoffman, E. (2017) "Project-Based Rental Assistance" In *Advocates' Guide*, National Low Income Housing Coalition: Washington, DC.

Jacoby, S. (2019) *Potential Flaws of Opportunity Zones Loom, as Do Risks of Large-Scale Tax Avoidance*. Center on Budget and Policy Priorities. Retrieved: 01.11.19.

Lane, S. (7.22.2021) "Housing Prices Hit a New High Up 23 percent in a year" In *The Hill.*

Layser, M.D. (2019) *A Typology of Place-Based Investment Tax Incentives,* 25 Washington and Lee J.Civ Rrs. & Soc. Just. 403.

McCabe, J. (2020) *Towards an Equitable Housing and Welfare Policy: The Past and Future Federal Government Intervention.* Thesis: Johns Hopkins University.

NLIHC (2021a) *The Problem.* www.nilhc.org/explore-issues/why-we-care/problem. Retrieved: 8.27.21.

NLIHC (2021b) *The Gap: A Shortage of Affordable Homes.* National Low Income Housing Coalition: Washington DC.

Ortiz, B. (2021) *Delivering Opportunity – Strategic Toolkit for Long-term Investment and Impact to Opportunity Zones in Springfield, MA.* Terminal Thesis: Clemson University Department of City Planning and Real Estate Development.

4 Transfer of Development Rights and Community Development

Evangeline Linkous

Overview

TDR is used throughout the world, with some notable examples in Taiwan, India, and Italy (Balakrishnan, 2019; Bjorkman, 2013; Chiodelli and Moroni, 2016; Shih and Chang, 2016). In the United States, there are more than 239 TDR programs, with the majority adopted in fast-growing states like Florida and California (Nelson, Pruetz, Woodruff, Juergensmeyer, and Witten, 2012). Perhaps, the best-known TDR program hails from King County, Washington. Recognized as the most successful TDR program in the United States in terms of the number of acres preserved, King County has protected 144,500 acres of rural and resource lands land in the Seattle region through transferred and banked development rights.

New York City is also well known for TDR due to its innovative and evolving use of the tool. TDR provisions are included in its 1916 zoning code—the first zoning code in the United States. In the 1960s, it adopted the tool for use in its historic preservation program, which was scrutinized in the landmark U.S. Supreme Court case *Penn Central Transportation Co. v. New York City* (1978). More recently, TDR in New York City made headlines when it was used as part of the High Line park development project and when the Hudson River Park Trust approved the sale of $52 million worth of development rights from publicly owned parkland.

Although TDR is widely used and continues to be a recommended strategy for achieving a variety of planning objectives, the performance of U.S. TDR programs is uneven. There are a handful of successful programs, but the vast majority of U.S. TDR programs fail to generate transactions and contribute to community planning objectives. Despite the discouraging track record, the planning research provides clear guidance on the factors that contribute to TDR program success (Pruetz and Standridge, 2008). When designed and administered in careful coordination with real estate market conditions, TDR is a tool that continues to offer promise for addressing a myriad of community development challenges.

DOI: 10.1201/9781003109679-5

TDR Program Design

TDR programs can serve three functions, which are often intermingled (Linkous, 2016). First, TDR supports the redistribution of development rights away from areas targeted for preservation and toward areas where additional development is appropriate. Second, TDR can be used to mitigate or compensate for property rights restrictions. For example, when zoning provisions are adopted that limit or reduce sending area landowner development rights—such as through downzoning—those landowners may be allowed to transfer those rights for economically viable use in another location. Third, TDR recruits private sector investment in support of community planning goals and ideally avoids negative fiscal impacts to the public sector—or even generates positive fiscal impacts.

The starting point for most TDR programs is the identification of an asset a community wants to protect from additional development, such as farmland, environmentally-sensitive lands, flood zones, or historic properties. These properties form the basis of a TDR "sending zone," or the area from which development rights may be transferred. The next step is the identification of a "receiving zone," where additional development rights may be transferred to. These typically include redevelopment areas where demand for additional density, building height, floor area, or other forms of development incentives are highly valued.

Many TDR programs are designed so that TDR sending and receiving areas correspond to zoning designations. For example, a sending area may encompass all land zoned for agriculture, and a receiving area may include all multi-family residential zones in the urban core. Alternately, sending and receiving areas may be linked to special planning districts established through zoning overlays or area plans. TDR programs that identify geographically distinct sending and receiving areas are referred to as dual-zone TDR programs.

Single-zone TDR programs, by contrast, usually allow for transfers within one defined area. These programs are designed to provide development flexibility while limiting aggregate development in accordance with community goals. New York City's Zoning Lot Mergers program uses TDR in this way—allowing property owners to shift development capacity between adjacent properties. This offers developers the opportunity to best utilize the development potential allowed under zoning and other development controls while still ensuring that zoning supports overall building height and bulk objectives for a given area. The Lake Tahoe Basin TDR program is a single-zone TDR program designed to limit overall development in the Lake Tahoe Basin to align with environmental capacities. A system of incentives encourages the transfer of development potential from environmentally sensitive areas with the Basin to those more suited for development. Single-zone TDR programs are also sometimes referred to as "floating" TDR programs. While they allow for development flexibility, these programs may create conflicts between conservation and development or may foster development inconsistent with community equity or urbanism goals (Linkous and Chapin, 2014; Shih, Chiang, and Change, 2019).

Pruetz and Standridge (2008) identified the ten factors associated with TDR program success, the most critical of which directly relate to ensuring that there is market demand for development rights. In other words, TDR credits must hold sufficient value that property owners will be willing to sell credits and developers will be willing to pay for them. Factors supporting market calibration include customizing receiving areas to best support growth, establishing strict sending area development regulations to limit development value, and providing few or no alternatives to TDR for achieving density bonuses in receiving areas. Daniels (2007) indicates that there should be twice as many receiving sites for TDR credits as there are TDR credits available from sending sites to ensure that credits are not oversupplied and will garner sufficient value.

A TDR bank is a central feature in many successful TDR programs. A TDR bank is an official entity authorized to buy, hold, and sell TDR credits. The bank can be administered by the local government or a non-profit third party such as a land trust. A TDR bank can offer certainty to both buyers and sellers that credits can be bought and sold, reducing transaction costs. TDR banks can also establish and stabilize TDR credit pricing. The most successful programs that use a TDR bank do so in coordination with an initial public investment—often supported by a bond issue—that stocks the bank with TDR credits available for purchase. This approach means that preservation goals are achieved upfront, and developers can readily purchase banked credits. The sale of credits can be used to pay down any debt incurred in the initial purchase or to create a revolving fund for future property acquisitions. This approach is used in King County, Washington and Palm Beach County, Florida, two of the most successful and long-running TDR programs in the nation.

TDR in Context: Case Studies in Community Development

TDR's popularity rests in large part on its ability to support a wide variety of community development objectives. This section provides case studies of three TDR programs that support divergent goals, but all contribute to community benefits and resiliency. Palm Beach County, Florida leverages TDR to preserve environmental areas while providing workforce and affordable housing. The Wynwood neighborhood in Miami, Florida maintains its vibrant cultural character in part through the use of TDR to preserve public art and open space. The City of Miami, Florida is looking to TDR to address the emerging challenges of climate mitigation and adaptation. Although all of these cases hail from Florida, they offer transferable lessons about the ways TDR can be tailored to meet distinct community needs and market conditions.

Workforce and Affordable Housing in Palm Beach County

Palm Beach County is part of the Miami metropolitan area and is home to approximately 1.5 million persons. While the eastern part of the county is highly urbanized, the western portion is home to important natural and agricultural areas just north of the Everglades and east of Lake Okeechobee. Following its first TDR ordinance, which passed in the 1980s and preserved 644 acres, the county updated its TDR program in the 1990s with provisions that continue to guide program operation today (Nelson et al., 2012).

The linchpin of Palm Beach's TDR program is its TDR bank, which the county stocked with 9,000 TDRs from environmental lands purchased through a $100-million bond approved by voters in 1991 (Nelson et al., 2012). In addition to allowing for the upfront preservation of 35,000 acres of the county's most critical natural areas, the bond-and-bank approach supports a revolving fund for ongoing maintenance and acquisition of environmental lands.

Under the current program, prices of TDR credits in the county bank are updated annually based on Palm Beach County median sales prices. Separate pricing is set for TDRs to be used for single-family and multi-family/condo development projects. Effective July 2021, single-family TDRs were priced at $44,000 (10% of the county's median single-family home price of $440,000) and multi-family/condo TDRs were priced at $24,500 (10% of county's median multi-family/condo home price of $245,000). Developers also have the option of purchasing TDRs directly from sending area landowners with whom prices can be negotiated.

Palm Beach County's TDR program supports the county's future land use policy of directing growth to existing urbanized areas in the eastern portion of the county while preserving the rural character out west. In general, eligible sending area lands include land zoned for rural residential, conservation, and agricultural uses, as well as environmentally sensitive lands designated for priority acquisition.

Receiving areas must be located within the county's designated Urban/Suburban Managed Growth Tier area (the designated urban growth area within unincorporated Palm Beach County) as part of Planned Development Districts, Traditional Development Districts, or residential subdivisions. The county uses a combination of tiered pricing and differentiated density bonuses to promote more density through TDR in desirable receiving area locations. Reduced prices are available for TDRs used as part of approved Neighborhood Plan projects (75% of full TDR price) or Revitalization, Redevelopment, and Infill Overlay projects (25% of full TDR price). Greater density bonuses are afforded to TDR receiving area projects east of Florida's Turnpike where urbanization is greatest, with additional bonuses for redevelopment projects or developments located near parks, transit, and employment facilities.

In response to the 2000–2005 Florida housing boom—which saw home prices in Florida rise by 82% (Rojas, McGuire, Ivey, and Durrenberger,

2007)—Palm Beach County adopted a voluntary Workforce Housing Program (WHP) in 2004 that aims to provide housing to the workforce the community relies on. The WHP program applies to new residential developments in the county's Urban/Suburban Tier, including both rental and for-sale units. The program targets four income categories:

- Low Income (< 60 to 80% of county median family income)
- Moderate 1 Income (< 80 to 100% of county median family income)
- Moderate 2 Income (< 100 to 120% of county median family income)
- Middle-Income (< 120 to 140% of county median family income)

Rental WHP units are required to target all four income categories, and for-sale units must target all but the Middle-Income category.

The WHP initiative works in tandem with Palm Beach County's TDR program to allow density bonuses to development projects that include WHP units. The county currently requires 34% of all TDR bonus density *rental* units and 29.75% of all TDR bonus density *for-sale* units to be WHP units. There is also a limited incentive option requiring just 17% of all TDR bonus density units to be WHP if these units target Low-Income and Moderate 1 households. Whereas TDRs for market-rate units are priced at 10% of Palm Beach County median housing sales prices, TDRs for WHP units are priced at 5%, with additional discounting for Neighborhood Plan, Revitalization, Redevelopment, and Infill Overlay projects. Developers may also opt for in-lieu fees or off-site provision of WHP units under certain conditions.

The WHP program is thoughtfully designed to be in sync with the marketplace, with linkages to both market supply and cost dynamics. However, the County has had to regularly adjust the program over the years in pursuit of better outcomes. Shortly after the program's adoption, housing prices began to cool and the economy subsequently plunged into recession, stalling development. As a result, the program did not perform as hoped, with just 62 WHP units approved or under review by March 2006 (Morse 2019). Although TDR units for WHP were provided at no cost or just one dollar during these early years, up to 50% of TDR bonus density was required to be WHP. In an attempt to be more responsive to market conditions, the county revised the program pricing and ratios in 2010 and 2019, landing on the current structure of 5% of full TDR pricing and a current maximum requirement of 34% WHP units. These changes reflect a strengthening economy and an attempt to increase developer willingness to provide on-site WHP units (rather than in-lieu or off-site) while also supporting the county's environmental fund (Reid, 2011).

In 2009, the county expanded its efforts to address housing affordability through TDR with introduction of the Affordable Housing Program (AHP), which serves households with incomes of less than 60% of area median income. The AHP program targets developments that provide a

minimum of 65% of the total number of dwelling units to qualifying households. TDRs for AHP projects are priced at 1% of the full TDR price and
can allow for a density bonus of up to 100% for projects that meet criteria
related to the geographic distribution of low-income housing across the
county as well as proximity to community facilities and services. The AHP
TDR program has generated only limited participation to date, but it is
just one component of the county's overall approach to affordable housing
provision.

As of this writing, the total number of TDRs sold under Palm Beach
County's TDR program is 2,540. The county indicated that development
outcomes for all of the TDRs could not be reported due to tracking mechanisms used, but provided data for the approximately 700 units for which
information was tracked, shown in Table 4.1 (Michael Howe, personal communication, February 2, 2021).

While the success of Palm Beach County's TDR program is evident in the
total number of TDRs sold, Table 4.1 shows that the TDR program is an
also important contributor in the provision of workforce housing in Palm
Beach County. Approximately 65% of TDRs tracked by use type were used
for the WHP program. TDR is also an important mechanism in the provision of market-rate multi-family housing in Palm Beach County. Although
not earmarked for specific income categories, such units are typically more
affordable compared to market-rate single-family homes. The county's
AHP program has not yielded a significant number of transactions, suggest-
ing that TDR may have limited value in supporting more affordable housing

Table 4.1 Palm Beach County TDRs Sold by Type

TDRs for Market-Rate Units	• 21%—Multi-Family TDRs • 7% Revitalization, Redevelopment, and Infill Overlay (RRIO) Multi-Family TDRs (25% of full TDR Price) • 4%—Single-Family TDRs • 2% Neighborhood Plan TDR Multi-Family (75% of full TDR price) • < 1% Countywide Community Revitalization Team Multi-Family TDRs (25% of full TDR price)
TDRs for the Workforce Housing Program (WHP)	• 27%—$1.00 TDRs • 33%—Multi-Family WHP TDRs • 3% RRIO Multi-Family WHP • 1%—Single-Family WHP TDRs • 1%—Countywide Community Revitalization Team TDRs (25% of full TDR price)
TDRs for the Affordable Housing Program (AHP)	• < 1%—Countywide Community Revitalization Team TDRs (25% of full TDR price)

Source: Palm Beach County (Michael Howe, personal communication, February 2, 2021).

that is significantly below the market rate. Overall, Palm Beach County's experience provides instructive lessons in tying TDR to market dynamics and leveraging demand for market-rate units to facilitate the provision of workforce housing.

Community Character in Wynwood

With its rich mix of galleries, restaurants, and retail establishments, the Wynwood neighborhood in Miami, Florida is world-class cultural destination distinguished by its industrial character and mural arts program. Originally developed a century ago as a manufacturing and logistics hub, Wynwood was once home to one of the largest garment districts in the United States. By the 1970s, deindustrialization and suburbanization contributed to growing crime, vacancies, and unemployment in the area. Revitalization of Wynwood took off in the late 1990s, prompted in large part by the work of real estate developer Tony Goldman, known for redevelopment work in New York City's Soho neighborhood, Philadelphia's 13th Street, and Miami's South Beach. Goldman envisioned the windowless warehouse and factory walls of Wynwood as canvasses for graffiti and street art. The Wynwood Walls, a curated open air mural art project, debuted in 2009 in coordination with Miami's annual Art Basel event. Since its inception, the program has invited more than 50 artists from around the world to cover over 80,000 square feet of walls (Wynwood Walls, 2021).

In 2013, the Wynwood Business Improvement District (BID) was formed to represent Wynwood area businesses. The BID "works to help the neighborhood achieve its full potential as a vibrant community for all that is centered on creativity, innovation, and art" (Wynwood Business Improvement District, 2021). In addition to supporting sanitation, safety, mobility, and marketing services, the Wynwood BID worked with City of Miami to create the city's first Neighborhood Revitalization District (NRD) plan in 2016. The NRD Plan establishes new zoning regulations for Wynwood to foster mixed-used developments, pedestrian-focused transportation, and preservation of the area's unique artistic and industrial character.

Wynwood's NRD Plan includes provisions establishing the Wynwood Transfer of Development Rights Program, the purpose of which is to encourage the establishment of Privately-Owned Public Open Space and preservation of Legacy Structures. Properties with either of these assets function as sending zones from which development rights may be transferred. Privately-Owned Public Open Space is defined as a publicly accessible area on a private lot. Legacy Structures are defined as existing buildings that are maintained and repurposed to contribute to the character of Wynwood. They must be associated with a significant industry important to the area's history, exemplify Wynwood's industrial past, or provide public art that is maintained in perpetuity. Although the public art component in Wynwood is typically murals, other art installations including 3-D and light art are also allowed.

Developments in Wynwood's high-density T5-0 and T6 Transect Zones may receive TDRs to enable additional building height. The program responds to the demand for development in Wynwood but also recognizes that the community's unique appeal comes in large part from the low-rise industrial buildings and public spaces that invite pedestrian activity.

Wynwood's TDR program is one of several TDR programs in the Miami-Dade area. The Wynwood TDR program is intentionally exclusive from the City of Miami's citywide TDR program for historic preservation, which has resulted in few transfers due to an oversupply of low-value credits. Wynwood's TDR program, which is relatively young, has generated two transfers to date with a third transfer in the pipeline. Transfer activity is anticipated to be somewhat limited because the TDR program competes with the Wynwood Public Benefits Program Wynwood, another NRD provision that allows bonus building height. The Public Benefits Program supports affordable and workforce housing, public parks and open space, civic space, and cross-block connectivity. City of Miami Planner Joseph Eisenberg (personal communication, October 15, 2020) explained that the Public Benefits Program offers greater flexibility to developers and thus appears to be a preferred option over TDR for attaining development bonuses (Figure 4.1).

Figure 4.1 In Wynwood, Miami, TDR helps enable a mix of protected low-rise and new construction high-rise development.

Source: Wynwood Business Improvement District (2019). Annual Report 2019.

TDR for Climate Adaptation and Mitigation

The application of TDR for climate adaptation and mitigation, sea level rise, recurrent coastal flooding, and managed retreat is only beginning to be explored. This section explores a set of emergent approaches to these issues and focuses on the work of the City of Miami to develop a TDR program for climate adaptation in its Arch Creek Basin Area. Before turning to the case study, the chapter offers guidance on the design of climate resiliency-focused TDR programs drawn from New York State's TDR enabling legislation, existing TDR programs for shoreline areas and water resources, and policy research.

New York State's TDR Enabling Legislation

The authority for local governments to enact TDR programs typically comes from state enabling legislation, although some local governments rely on powers delegated through home rule. Approximately half of U.S. states have adopted TDR enabling legislation, which typically includes three main principles (Nelson et al., 2012). Foremost, enabling legislation authorizes conveyance of development rights while also providing clarity about the granting of authority, instruments to be used for transfers, and transfer recording procedures. Second, enabling legislation establishes standards for voluntary participation in TDR transfers, typically with guidance about which parties must give consent. Finally, state enabling legislation usually identifies the resources that TDR can be used to preserve.

Although New York City has long used TDR, New York State only formally enabled TDR in 1980, when it authorized the use of the tool for historic preservation. In 1989, the State amended its Town, Village, and General City Law to allow the use of TDR to protect agricultural lands and open space of special historical, cultural, aesthetic, or economic interest. In 2019, following the devastating impacts of Hurricane Sandy, the State Legislature amended its TDR enabling legislation to allow the use of TDR to protect lands at risk from sea level rise, storm surge, or flooding.

New York's TDR enabling legislation offers a model for states seeking to apply TDR for climate mitigation and adaption. Beyond explicitly allowing for use of TDR in this context, the legislation addresses the use of TDR for land that may be submerged, shifting, or otherwise dynamically impacted by climate change. States considering similar updates to TDR enabling legislation should assess how their rules on TDR authority, conveyance, and consent should be amended in light of the conditions associated with climate change.

Coastal, Shoreline, and Water Resource TDR Programs

Although TDR programs specifically for climate mitigation and adaptation are only beginning to emerge, TDR has long been used to manage environmentally-sensitive coastal and shoreline areas, as well as to protect other water resources (such as wetlands or groundwater) or attributes (like water quality). Such programs offer insights that may inform contemporary TDR programs aimed at climate resiliency.

Many of the TDR programs designed for environmentally sensitive water resources are single-zone TDR programs that identify a single area within which transfers may occur. In these programs, TDR provides incentives to cluster development, limiting impervious surfaces and preventing build out of sensitive areas. Monroe County, Florida provides an instructive example of a single-zone TDR program for coastal area management. All of Monroe County, which is home to the Florida Keys and portions of Everglades National Park, is located within the 100-year floodplain. In Monroe's program, which dates to 1986, residential or hotel density can be transferred between any two sites in the county as long as the sending site has greater ecological signif-icance (as identified through the land use classification) and equal-or-lower permitted density than the proposed receiving site. A suite of complemen-tary regulations for environmentally-sensitive lands—such as a requirement for 85% open space—strictly constrains development in sensitive areas and increases the appeal of TDR (Pruetz, 2003). Monroe's TDR program shows slow but steady participation, with more than 100 credits transferred.

A few communities use a dual-zone TDR program design to support coastal area protections. One example is Brevard County, Florida, located east of Orlando along Florida's Atlantic Coast. Brevard's TDR program was adopted in 1979 in tandem with downzoning provisions for coastal areas. In Brevard's program, sending areas may include oceanfront properties or land zoned as environmental or agricultural areas, and receiving areas are comprised of multi-family residential zones. Brevard's program has failed to generate transfers, in part because the market for multi-family residen-tial is weak, and desired densities can usually be achieved through other means (Pruetz, 2003). Moreover, oceanfront property owners prefer to build homes rather than relinquish rights. Oxnard, California presents a similar case. Oxnard is located west of Los Angeles on the Pacific Coast. Oxnard adopted a TDR program in 1984 to preserve environmental resources, max-imize coastal access and recreation, and limit development in potentially hazardous areas. All oceanfront properties are designated as sending sites, and receiving sites are in designated multi-family residential zones (Pruetz, 2003). The program has generated just a few transfers due to landowner preference to develop beachfront lots.

The experiences of these dual-zone TDR programs suggest that the unique and highly valued locational attributes of oceanfront and shoreline prop-erties create marketplace challenges for TDR. Although dual-zone TDR programs create clear rules about where growth is appropriate or should be limited, the personal and monetary value placed on waterfront homes makes commodification of development rights from such areas a significant chal-lenge. For climate mitigation and adaptation purposes, single-zone TDR programs may offer more opportunities because they do accommodate some development in desirable places near water bodies. These programs have better track records but can lead to conflicts between conservation and development (Linkous and Chapin, 2014). Furthermore, single-zone TDR programs may not afford an appropriate solution for areas where resiliency planning necessitates the full removal of development rights across an area.

Recognizing that shoreside living is a lifestyle choice that homeowners will be reluctant to give up, Lung and Killius (2016) offer several solutions to address the challenge of designing TDR programs that offer sufficient compensation to prompt supply-side property owners to participate. First, impact fees could be levied on areas prone to flooding where emergency services and mitigation measures are required. Such fees would help defray these expenses and could also stimulate the supply for development rights by making homeowners accountable for these costs. A second approach is to structure TDR like a corporate tender offer. In this model, a local government establishes a finite number of transfers with a set level of compensation that is made available to homeowners facing flood loss. Homeowners are incentivized to take part before the set number of rights are tendered. In addition, a tiered tender approach could allow for more generous compensation to early adopters, encouraging homeowners to transfer sooner. Third, for low-lying coastal areas, TDR could be linked to wetland mitigation credits; this option would allow the conversion of TDRs into wetland mitigation credits, which could be purchased from the local government if banked. Lung and Killius also recommend including a provision for deferred taxes to avoid any actual or perceived penalty while the TDR transaction is in process.

Arch Creek Basin Area

With 84 coastal miles, 2.7 million residents, and nearly half a trillion in at-risk assets, the City of Miami, Florida is widely recognized as among the most climate-vulnerable areas in the U.S. Among its responses to these challenges, Miami is planning for the county's Arch Creek Basin Area, a 2,838-acre stormwater drainage basin area located in the northeastern part of the county along Biscayne Bay. Approximately 67% of the Arch Creek Basin Area is located in a Special Flood Hazard Area. The area is economically diverse and densely populated, with much of the housing stock built prior to the establishment of the National Flood Insurance Program and before county flood regulations were adopted. With many parts of Miami-Dade County facing vulnerabilities similar to those in the Arch Creek Basin Area, strategies used here may have broad applicability for future adaptation initiatives.

In 2016, the Arch Creek Basin was selected as the City of Miami's first Adaptation Action Area (AAA). The AAA program is an optional planning designation first authorized by the State of Florida in 2011 to support planning, technical assistance, and funding opportunities for areas vulnerable

What is a city slough?

"A city slough park concept would create green space within a neighborhood to manage floodwater"

ULI Arch Creek Basin Advisory Services Panel finding, May 2016

IMAGE SOURCE: WALTER MEYER, ULI

These images illustrate the short-range, mid-range and long-range implementation of a city slough concept discussed and endorsed by the 2016 Resilient Redesign Workshop and the ULI Arch Creek Basin Advisory Services Panel Report.

Figure 4.2 In Miami's Arch Creek Basin Area, a TDR proposal would remove flood-threatened homes from harm's way and to create a new city slough park.

Source: Urban Land Institute (2017).

to climate change. To inform the implementation of the AAA in the Arch Creek Basin, Miami recruited the Urban Land Institute's (ULI) Technical Assistance Panels service. The final report for ULI's study was produced in May 2016 and identified both broad and specific recommendations, including the development of a TDR program to support the implementation of a new, large-scale slough park in the Arch Creek Basin that would transform the area's historic creek corridor into a neighborhood green space, as shown in Figure 4.2. In addition to serving as a floodwater management tool, the slough park would create new recreational areas and serve as a focal point for future development (Urban Land Institute, 2016). Parcels to be included in the park could comprise the sending area, and the North Miami Community Redevelopment Area was suggested as a potential receiving area.

In 2017, ULI joined with the Miami-Dade County Office of Resiliency to form a TDR Focus Group to further explore the use of TDR for the Arch Creek Basin. The group's report (Urban Land Institute, 2017) includes several recommendations, including identifying focused sending and receiving area sites, ensuring careful management of supply and demand of TDR credits, creation of a TDR bank, and use of dwelling units as the basis for transfers. The latter recommendation is designed to create a competitive market given that the city's historic preservation TDR program, which is based on floor area ratio transfers, is oversaturated. Recognizing that the sending area will primarily be made up of single-family homes with little overall density available to transfer, the report recommends a multiplier of five that would allow one unit of TDRs transferred from the sending area to enable five units of development in the receiving area. The report notes that, unlike some TDR programs where sending area conservation easements allow owners to retain some property rights, the Arch Creek Basin TDR initiative is designed to convert sending area properties into a publicly-owned slough park and thus aims at a total buyout of properties. Finally, the report recommends allowing for a time separation between the contract and closing of sending area property transactions. This provision allows residents to remain in place at least until the next major flood event. Together, these provisions aim to provide the certainty, flexibility, and value that could support voluntary relocation of residents out of flood-prone areas.

The report provides detailed guidance about the benefits of a TDR bank for TDR programs aimed at climate adaptation. In addition to the typical benefits of a TDR bank—including stabilizing prices and reducing transaction costs—a TDR bank can offer unique opportunities in a flood risk context. For example, the report indicates that a banking program would allow sending area property owners the opportunity to pre-sell a property. The TDR bank could offer the credit for sale to support the purchase of the property, but the homeowner could remain in place for up to ten years or until inundated. Banked properties could only be renovated up to a certain percentage of value and would need to be vacated and sold if damaged to a pre-determined threshold. While the property is banked, eligible occupants could be given priority for low-income housing options. Additionally, such a program could also allow the local government to act as an interim buyer for FEMA buyout transactions (Urban Land Institute, 2017). A final recommendation of the report is to link the program to a nearby low-income housing tax credit project. This equity-focused solution recognizes that many flood-prone areas are home to low-income residents who will benefit from retaining community connections. Miami-Dade's efforts to adapt TDR to a climate adaptation context provide helpful guidance on program technical design but also point to the exciting opportunities to link TDR to broad community benefits.

Conclusion

TDR programs provide development entitlements or density bonuses in exchange for public benefits—usually in the form of land or historic property preservation. As the examples in this chapter show, the tool is highly customizable and can be readily leveraged to support a wide range of community development goals including housing affordability, preservation of community character, and climate adaptation and mitigation.

Although TDR offers unique opportunities, it also presents exceptional challenges that are especially relevant in the community development context. Historically, TDR programs in the United States have an uneven track record, and this is largely due to the way TDR requires developers to pay for the development entitlements or density bonuses TDR enables. If these can be procured through other means, such as through by-right or amended zoning, TDR programs cannot function. TDR programs must be designed with a market-driven perspective, with attention paid to the nexus between the cost associated with achieving preservation and community benefits in sending areas (typically, the price at which a private property owner will be willing to sell property rights) and the return a developer can anticipate from new development in receiving areas.

Although all TDR programs are designed to generate a public benefit, TDR programs aimed at community development face the additional challenge of putting community objectives at the forefront. Making these community benefits a reality within the market-driven constraints of TDR create real program design challenges. However, the case studies in this chapter show that there are viable program design solutions aimed at improving property owner and developer participation. Palm Beach County's strategy of tying TDR pricing to median home sales prices reflects a program design strategy that is responsive to evolving market cycles. This approach allows the county to take advantage of robust market conditions to capture increased community benefits in the form of workforce and affordable housing as well as open space protections. Wynwood's TDR program addresses the niche needs of a global cultural hub. Its TDR program leverages development demand to protect historic properties, urban open space, and street art. Although Wynwood's TDR program has generated limited participation, it was intentionally designed to be just one among a host of options that support community development objectives—providing flexibility to ensure the right mix of incentives is available for different projects. Finally, the innovative thinking around the use of TDR for flooding and sea level rise applications grapples with a mix of issues including climate gentrification and the combination of short-term (flood-related) and long-term (sea level rise) time frames that shape coastal area planning. Miami's efforts to protect and re-house vulnerable residents, manage floodwater, and create a new park in the Arch Creek Basin are at the forefront of emerging TDR program designs that link social, economic, and environmental resiliency. As these examples show, TDR is a tool with potential that continues to be reimagined to address contemporary planning challenges.

References

Balakrishnan, S. (2019). *Shareholder Cities: Land Transformations Along Urban Corridors in India*. Philadelphia PA: University of Pennsylvania Press.

Bjorkman, L. (2013). Becoming a Slum: From Municipal Colony to Illegal Settlement in Liberalization-Era Mumbai. *International Journal of Urban and Regional Research* 38(1), 36–59.

Chen, H. (2019). Cashing in on the Sky: Financialization and Urban Air Rights in the Taipei Metropolitan Area. *Regional Studies* 54(2), 198–208.

Chiodelli, F. & Moroni, S. (2016). Zoning-Integrative and Zoning-Alternative Transferable Development Rights: Compensation, Equity, Efficiency. *Land Use Policy* 52, 422–429.

City of Miami. (December, 2020). *Climate Change in the City of Miami.* City of Miami. https://www.miamigov.com/Government/ClimateChange

Linkous, E. (2016). Transfer of Development Rights in Theory and Practice: The Restructuring of TDR to Incentivize Development. *Land Use Policy* 51, 162–171.

Linkous, E. & Chapin, T. (2014). TDR Program Performance in Florida. *Journal of the American Planning Association* 80(3), 253–267.

Lung, J. & Killius, M. (2016). Tools for a Resilient Virginia Coast: Designing a Successful TDR Program for Virginia's Middle Peninsula. *Virginia Coastal Policy Center.* https://scholarship.law.wm.edu/vcpclinic/25

Nelson, A., Pruetz, R., Woodruff, D., Nicholas, J., Juergensmeyer, J., & Witten, J. (2012). *The TDR Handbook: Developing and Implementing Transfer of Development Rights Programs.* Washington DC: Island Press.

Pruetz, R. (2003). *Beyond Takings and Givings: Saving Natural Areas, Farmland, and Historic Landmarks with Transfer of Development Rights and Density Transfer Charges.* New York: Arje Press.

Pruetz, R. & Standridge, N. (2008). What makes Transfer of Development Rights Work?: Success Factors from Research and Practice. *Journal of the American Planning Association* 5(2), 78–87.

Reid, A. (2011, January 27). Palm Beach County Lowers Prices for Developers to Build More Homes than Usually Allowed. *Sun Sentinel.* https://www.sun-sentinel.com/news/fl-xpm-2011-01-27-fl-developer-perk-palm-20110127-story.html

Rojas, G., McGuire, S., Ivey, S., & Durrenberger, T. (2007). The Florida Housing Boom. *Florida Focus* 3(2), 1–16.

Shih, M. & Chang, H. (2016). Transfer of Development Rights and Public Facility Planning in Taiwan: An Examination of Local Adaptation and Spatial Impact. *Urban Studies* 53(6), 1244–1260.

Shih, M., Chiang, Y., & Chang, H. (2019). Where does Floating TDR Land? An Analysis of Location Attributes in Real Estate Development in Taiwan. *Land Use Policy* 82, 832–840.

Urban Land Institute. (2016). *Arch Creek Basin, Miami-Dade County, Florida.* Washington DC: ULI Advisory Services Panel Report.

Urban Land Institute. (2017). *Exploring Transfer of Development Rights as a Possible Climate Adaptation Strategy.* Urban Land Institute Resilience Panel Focus Group with Miami-Dade County.

Wynwood Business Improvement District. (2021, January). *Our Story.* Wynwood Business Improvement District. https://wynwoodmiami.com/learn/our-story/

Wynwood Walls. (2021, January). *About Wynwood Walls.* Wynwood Walls. https://thewynwoodwalls.com/overview

5 Historic Building Reuse as a Form of Community Real Estate Development

Barry L. Stiefel

Introduction

Besides being the most in-demand, or valued real estate site locations in the world, what do the Dome of the Rock on the Temple Mount in Jerusalem, Israel; Buckingham Palace in London, United Kingdom; Eiffel Tower, in Paris, France; Taj Mahal, India; the White House in Washington, D.C.; Independence Hall in Philadelphia; Upper Fifth Avenue in New York between 50th and 60th streets; Bond Street, London; and the Avenue des Champs-Élysées in Paris, all have in common? They are *all* also historic places. While there are a scattering of contemporary buildings along Upper Fifth Avenue, Bond Street, and Avenue des Champs-Élysées, many (and sometimes most) are old. For many real estate developers, having the opportunity to own, manage, and work on or in the vicinity of one of the previously mentioned property locations is a signifier that they are extremely successful and accomplished. However, if you (hypothetically or not) possessed one of these highly esteemed properties, what would you do with it? Would you tear it down and rebuild something bolder and grandeur? Or would possessing the property as a responsible custodian for stewarding it in a historically-sensitive manner until the next guardian be sufficient enough for your investment portfolio and ego? After all, all of these properties are historic, and short of an unplanned catastrophe, will continue as such.[1]

Discussing the future and fate of high-end properties such as Notre Dame Cathedral or Independence Hall, among others from this list in respect to real estate development seems outrageous, especially in regard to a community-oriented perspective, but what separates the two approaches of community real estate development from conventional is sometimes only a matter of perspective and time. A place that first matters to a community can become significant to a nation if the argument that says as much resonates sufficiently among the entire people. Indeed, one of the first historic preservation initiatives in the United States began in this very way with Independence Hall in 1816. In 1813, the Pennsylvania state government decided that the building – then called the Pennsylvania State House, originally built by the British colonial administration in 1732 – could be demolished in

DOI: 10.1201/9781003109679-6

order to make way for a modern building. In response, many Philadelphians organized a grassroots effort to prevent the destruction of the Pennsylvania State House by buying the property and surrounding lots.[2] Some other 18th century buildings around Independence Hall were also saved, though others – such as the President's House where George Washington and his enslaved housekeepers, and John Adams had lived before there was a White House in Washington, D.C. – were destroyed for new development. Because Independence Hall mattered to local Philadelphians it was protected, from which the entire nation now benefits. The President's House did not matter to Philadelphians during the early 19th century, which is why during the 2000s, the National Park Service and a combination of other public and private sources spent more than $10.5 million to do archaeological work and an interpretative exhibit on the site that features Washington's enslaved housekeepers who were once there, since the Black presence at Independence Hall has otherwise been erased, which mattered to a national audience.[3] Thus, Independence Hall evolved from a historic building where concern for it expanded from the local to the national level because the historic significance was "rediscovered" by the rest of the country after enough time had lapsed so that it had accrued a sufficient level of age value to be properly appreciated.

The focus of this chapter will address historical building reuse as relevant to community real estate development, though understanding that the perception of what makes historic properties of local, regional, national, or even international importance can change and evolve over time. Many communities that were once considered a low development priority are now being rediscovered for their re-developable urban cores, such as the example that historic Philadelphia presented with Independence Hall during the early 19th century. So, formerly undesirable historic buildings are now in demand. Historic redevelopments – often called rehabilitation or adaptive reuse – often have higher standards and are, at times, not practical for conventional developers who are habituated to operating with greenfield or greyfield properties where there are fewer restrictions. Thus, this chapter will examine the use of different legal aspects, financial incentives, and a human-centric approach that encourages community-driven historic real estate rehabilitation.

Previous Studies on Historic Properties and Real Estate Development

One of the more respected authorities of historic reuse as a form of community real estate development is Donovan D. Rypkema, principal of the Washington, D.C.-based real estate and economic development-consulting firm, PlaceEconomics. Rypkema's work has been both geographically and topically vast within the United States. Among his assistance-oriented publications with the National Trust for Historic Preservation are *Feasibility*

Assessment Manual for Reusing Historic Buildings, The Economics of Historic Preservation: A Community Leader's Guide, A Guide to Tax-advantaged Rehabilitation, Community Initiated Development: A manual for non-profit real estate development in traditional commercial districts, and *Historic Preservation and Affordable Housing: The Missed Connection.*[4] Rypkema's manuals and guidebooks are excellent places for community-oriented real estate developers, planners, and property managers unfamiliar with historic buildings to begin learning about the special opportunities that are available and the unique challenges associated with them.

Other works, such as *Landmarks Preservation & the Property Tax: Assessing Landmark Buildings for Real Taxation Purposes* by David Listokin, and Judith Reynold's *Historic Properties: Preservation and the Valuation Process* are also useful resources for better understanding the property tax liabilities and real estate value of historic buildings with special characteristics that make ascertaining their assessment more complicated than conventional real estate. Listokin specifically looks at how officially designated historic properties have been complemented by increasing interest in other incentive and disincentive government-driven programs, as well as how property tax assessment can become an influential catalyst or disadvantage for building reuse, depending on the circumstances.[5] Reynold provides an informative description of Real Estate Valuation Theory that is specific to historic properties, with applications of the Sales Comparison, Cost, and Income approaches as well as how to best reconcile Market Value indications that incorporate the unique financial peculiarities of old buildings that conventional property appraisers might not always consider.[6] For those interested in more holistic projects designed at preserving and integrating old buildings and cultural heritage into community development approaches, Guido Licciardi and Rana Amirtahmasebi co-edited *The Economics of Uniqueness: Investing in Historic City Cores and Cultural Heritage Assets for Sustainable Development.* Licciardi and Amirtahmasebi take an international approach for organizing investment in historic buildings and neighborhoods in such a manner that enhances job creation and economic welfare for local residents, especially the poor, while also enhancing inclusive growth and sustainable urbanization.[7]

In recent years, there has emerged intersectional interest between community real estate development, human-centered design, and sustainability[8]; this has also been the case with historic preservation and building reuse. Human-centered design is a method for solving problems with solutions coming from the human perspective in all steps of the process, and sustainability is about meeting current needs without inhibiting future generations from meeting their own needs. *Community-Built: Art, Construction, Preservation, and Place,* co-edited by Katherine Melcher, Kristin Faurest, and myself look at how

> community members have come together to build places… illustrate how
> the process of local involvement in adapting, building, and preserving a

built environment can strengthen communities and create places that are intimately tied to local needs, culture, and community. The lessons learned from this volume can provide community planners, [community-oriented real estate developers,] grassroots facilitators, and participants with an understanding of what can lead to successful community-built art, construction, preservation, and placemaking.[9]

Continuing with this vignette of thought, Jeremy C. Wells and I co-edited *Human-Centered Built Environment Heritage Preservation: Theory and Evidence-Based Practice* to address "the question of how a human-centered [preservation] approach can and should change practice ... to [better] manage cultural landscapes, assess historical significance and inform the treatment of building and landscape fabric... this approach is essential for creating an emancipated"[10] community real estate development practice that successfully engages different perspectives. Lastly, *Sustainable Heritage: Merging Environmental Conservation and Historic Preservation*, co-authored by Amalia Leifeste and I provides

> a holistic critique of the challenges we face in light of climate and cultural changes occurring from the local to the global level. It synthesizes the best practices offered by separate disciplines as one cohesive way forward toward sustainable design,[11]

which is needed in community real estate development. This book considers "strategies for increasing the physical and cultural longevity of the built environment, why these two are so closely paired, and the potential their overlap offers for sustained and meaningful inhabitation,"[12] which is essential for community-minded developers interested in sustainability at historic places.

Knowing What To Do and What You Have

Many have tried real estate development but do not always succeed. In *Last Harvest: From Cornfield to New Town*, the author Witold Rybczynski describes how while George Washington was ultimately successful as a military General leading American rebels in independence from the British and serving as the first President of the United States, he did not do well as a real estate investor and developer. Knowing what to do and what kind of property you have is essential in the real estate business.[13] Ian Formigle, Vice President of Investments at CrowdStreet, a real estate investing platform, and author of *The Comprehensive Guide to Commercial Real Estate Investing*, has articulated a highly structured Real Estate Development Process, with the three stages of Pre-Development, Construction, and Operation divided into several smaller, incremental steps.[14] The U.S. Department of Housing and Urban Development also describes a Community Real Estate Development process that entails Forming the Development Concept,

Feasibility Study, Deal Making: Planning and Financing, Project Construction, and Operation or Sale.[15] Others, such as *Property Like a Pro*, provide their own critical step guides.[16] Traditional developers are often the ones who contribute equity to projects, own the finished products, and receive the financial returns the projects ultimately generate, whereas community-based developers may or may not do the same because sometimes they are motivated by other goals that are community-oriented and at other times they have different equity partners (such as local governments, land trusts, etc.).[17] Community-based developers do many of the same projects as conventional developers, it is just that the return on investment is different and a different cohort of stakeholders may be benefiting from it. In either case, the developer may still have complete control of the asset whether that is a stand-alone project or is part of a larger development. However, in a brief survey of these recommended real estate development processes and guides, no mention was found that specifically mentioned the complexities of historic building reuse that go beyond the financial advantages, legal aspects, history, or aesthetics of a place that make it special. For example, an important step before undertaking physical work for any historic building reuse project is to study and understand the building and surrounding site, which is done through research and writing of a historic structures report and is used as a preservation planning tool to help guide the project. Besides documenting and providing information about a property's existing conditions and history, the report also addresses management and development opportunities for the building's reuse. The National Park Service has developed an informational brief on *The Preparation and Use of Historic Structure Reports*, which can help community real estate developers understand the purpose of this planning tool in greater detail, but specialized professional expertise should be recruited for authoring such studies.[18]

One common public misconception is the strictness of historic preservation-related laws. While National Historic Landmarks (the most prestigious historic designation level on the spectrum) and properties on the National Register of Historic Places are held in higher regard than those on a State Register of Historic Places or in a local municipal landmark program, the regulations that provide protection to these buildings being reused are often inversely restrictive. In other words, the ordinances and building codes that regulate local municipal landmarks are the strictest and the federal laws for nationally recognized historic buildings the least strict. This is exemplified most with places on the World Heritage List, the most prestigious recognition for a historic property through a program run by UNESCO, where there are *no* protections provided by this designation other than shameful public publicity if one purposely causes harm to such a place. However, builders on the National Register of Historic Places are often automatically put on local municipal landmark lists so that these properties also receive stricter protections. In regard to the American legal tradition, the U.S. Supreme Court ruled in the landmark case *Penn Central Transportation Co. v. City of*

New York, 438 U.S. 104 (1978) that historic preservation is "an entirely permissible governmental goal," which most state and local governments conduct through laws, building codes, and zoning ordinances.[19] Indeed, some states have a separate historic building code from their regular one, which offers alternative building regulations for conducting repairs, additions, and modifications needed for the adaptive reuse, preservation, and continued use of historic buildings that have been recognized through a government-recognized program, such as the National Register of Historic Places, State Register of Places (though not all states have this), or municipal landmarks list. The National Park Service also provides special informational assistance for owners of historic buildings in becoming compliant with the Americans with Disabilities Act, in *Making Historic Properties Accessible.*[20] Cities from Charleston, South Carolina to Las Vegas, Nevada, as well as county (boroughs and parishes) governments for unincorporated areas, have historic district overlay zoning that is generally guided by the Certified Local Government program. So, when a developer or other property owner encounters historic property zoning regulations that they believe are overly stringent, these ordinances should be understood as a reflection of the majority of voters who live within that area who would like the aesthetics to be maintained as such, as the U.S. Supreme Court upheld in *Berman v. Parker*, 348 U.S. 26 (1954). This situation is inclusive of vacant lots within or next to an established historic district because new development is also regulated to be aesthetically compatible.

Assuming that local ordinances do not violate a federal or state "takings clause" or other laws, it is municipal and county councils elected through a popular vote that has the authority to ratify, change, and repeal historic property overlay zoning. Community real estate developers need to be mindful of this democratic process so that communities are maintained and strengthened.[21] However, developers should also be aware that while historic district overlay zoning may restrict some usages of property and thus are assuming an increase in business risk; empirical studies from multiple cities have found that in comparative instances where similar properties were located both within and outside of a community's historic district(s), and all other factors were relatively equal, that the real estate inside of the historic district sold for higher prices. The historic district zoning was the only factor that could be attributed for the higher market price of the property. The rationale is that preservation restriction can outweigh the negative effects on development because the real estate market is assured that protected historic properties are more likely to endure intact longer into the future than those that are not certified historic. On the flip side, municipal governments should also be aware of developer interests and that consistency in historic district overlay zoning over time and geography (within city limits as well as with neighboring municipalities) are needed in order to have a healthy historic property real estate market so that property owners are not adversely surprised by inconsistencies.[22]

Financial Opportunities for Reusing Historic Buildings

While local zoning ordinances and other building code regulations are often more controlling than state and federal regulations on what is or is not permitted to historic properties, it is at the higher levels of government where there are greater financial inducements for historic building reuse, such as tax credits and easements. However, through utilizing higher level government historic preservation financial opportunities, one is also accepting that they are adopting stricter expectations for project design and development through the Secretary of the Interior's Standards for the Treatment of Historic Properties. These Standards were established to encourage historic preservation best practices and are subdivided into Preservation, Rehabilitation, Restoration, and Reconstruction methodologies to reflect the most common project approaches.[23] For community real estate developers, the Secretary of the Interior's Standards for Rehabilitation are the most commonly used because of their required utilization in building reuse tax credit projects, though Preservation may occasionally be used in the instance of building façade conservation easements. In many instances, state and local governments use the Secretary of the Interior's Standards for the Treatment of Historic Properties as their rulebook for proper historic building stewardship. The Secretary of the Interior has also drafted additional guidelines for Sustainability (weatherization, insulation, HVAC, solar technology, wind power, cool/green roofs, etc.), Flood Adaptation (dry floodproofing, wet floodproofing, basement fill, elevating on new foundation, relocation, etc.), and Cultural Landscapes (Environmental Protections, etc.) to further assist property developers in rehabilitation projects trying to go green and/or prepare for climate change while they also pursuing historic preservation financial opportunities.[24]

The Federal Historic Preservation Tax Incentives program is the largest and most popular of the financial opportunities, which were first created by the 1976 Tax Act. Prior to this, not only were there no historic preservation tax credits but the tax code actually favored the demolition of old buildings to promote their replacement with new construction. To qualify for the 20% income tax credit program, an income-producing old building that is "certified historic" (either on the National Register of Historic Places or a contributing property within a National Register Historic District) must follow the Secretary of the Interior's Standards for Rehabilitation. State Historic Preservation Offices are the first to inquire for support and information for developers interested in the 20% income tax credit program, with the National Park Service assisting in the administration of the program in cooperation with the Internal Revenue Service. Non-income-producing buildings, such as owner-occupied residential housing, do not qualify for this federal tax credit program. Depending on state regulations, developers working on qualified rehabilitation projects who are not in a position to use the full 20% income tax credit can offer the credits to third parties, such as a bank or other financial institution, to generate additional capital for the project and thus decrease the quantity of debt financing needed for the development project. Qualified rehabilitation expenses include "costs related to the repair or replacement of walls, floors, ceilings, windows, doors, air conditioning/heating, systems, kitchen cabinets and appliances, plumbing and electrical fixtures, architects' fees, construction loan interest, and environmental reports"[25] within the historical sections of the property. The financial value of the tax deduction is tabulated by multiplying the amount of qualified rehabilitation expenses costs by the 20% tax credit rate. On average, the National Park Service's Technical Preservation Services approves more than 1,200 projects annually which encourages more than $6 billion yearly in the private investment of old buildings. There used to be a 10% income tax credit program for "non-certified historic" buildings constructed before 1936, but this was repealed in 2017.[26]

The Revenue Act of 1978 made an additional historic preservation financial opportunity through the creation of tax deductible conservation easements on buildings, in addition to ones available on vacant land. The donation of an easement permits the property owner to retain use and title of a historic building while also safeguarding its long-term preservation along with a one-time tax deduction. A historic preservation easement (usually part of a building's exterior or other historically significant elements) is a special type of conservation easement that entails a legal contract with a power invested in a qualified non-profit private historic preservation organization or government agency, where the owner consents to maintain the building according to one of the Secretary of the Interior's Standards for the Treatment of Historic Properties. Committing to these treatment standards safeguards that the building's integrity, value, and historic context are maintained. Easements are most often granted in perpetuity, where they are filed with the county land records, binding the current and future owners to the easement contract. The owner maintains a major interest in the building and can sell or transfer it to whomever they wish. Since most buildings are unique, easements contracts are individually written for each property's special circumstances regarding the specific elements to be preserved. Following the completion of the historic preservation easement gift, the receiving qualified non-profit private historic preservation organization or government agency provides the owner with a receipt of the easement's appraised property value which can be used as an income tax deduction.[27] Depending on the locality, the property's assessed value may have decreased slightly because of the easement gifting, so the owner could also have the building reassessed, which will likely be a lower amount in these circumstances. All certified historic properties owned by taxable entities qualify for the historic preservation easement program, including owner-occupied housing. Property owners with tax-exempt status can still engage in historic preservation easement contracts but they do not have a tax liability to use the deductible gift receipt donation for.

The 20% historic rehabilitation tax credit and historic preservation ease-ments are the two financial opportunities that are specific to historic pres-ervation, though depending on the kind of project, other building reuse projects may qualify for additional financial assistance programs. Those of interest to community real estate developers include the New Market Tax Credit and Low-Income Housing Tax Credit programs. The New Market Tax Credit program is for enticing private investment into low-income com-munities by allowing corporate and individual investors to obtain a tax credit against their federal income tax in exchange for financially investing in specialized Community Development Entities. Recipients of qualifying projects receive a 39% credit of the original investment value and can be claimed over a seven year period.[28] The Low-Income Housing Tax Credit program provides designated local and state allocating government agen-cies approximately $8 billion annually to issue tax credits for the obtain-ing, adaptive reuse, or new development of rental housing for lower-income households.[29] Projects for the New Market Tax Credit and Low-Income Housing Tax Credit can take place at historic buildings that are being re-used, as well combined with the previously mentioned 20% historic preserva-tion rehabilitation tax credit and easement opportunities. Moreover, when working within public-private partnerships, community real estate develop-ment projects may also be eligible for the federal government's various grant programs, including Community Development Block, Affordable Housing Development and Preservation, Community and Economic Development, Environment and Energy, Fair Housing, Homelessness, Homeownership, Rental Assistance, Supportive Housing and Services, Intermodal Surface Transportation Efficiency, among others.[30]

Besides federal tax credit and grant programs, there are also many op-portunities that exist at the state and local levels, depending on the juris-diction. Multiple states, such as South Carolina, have their own historic preservation rehabilitation tax credit that mirrors the federal government's 20% program. In the instance of South Carolina, their state historic preser-vation rehabilitation tax credit can also be applied to owner-occupied resi-dences in addition to income-producing building. South Carolina also has a Special Property Tax Assessment program to assist property owners in fixing up their historic buildings.[31] Other states have their own unique his-toric preservation-related financial opportunity programs, such as Califor-nia with their Mills Act tax abatement program done in coordination with local governments, which is in addition to their own state historic preser-vation rehabilitation tax credit program.[32] While not historic preservation-focused, state Enterprise Zone programs, such as in Ohio, can be used as an economic development tool for partial personal and real property tax exemptions for businesses located in historic building that is being reused.[33] Other states, such as Illinois, have Tax Increment Financing districts that can be overlaid with historic neighborhoods so that public financing can be used as a subsidy for community improvement and redevelopment projects.

Illinois's Tax Increment Financing districts can be used to redevelop old or vacant buildings, fund public infrastructure improvements, clean up pollution, and assist the viability of central business districts.[34]

Several cities in the United States have their own historic preservation-related financial opportunities and advantages too. A common one, such as in some California cities, is streamlining the permitting and approval process for real estate development projects that meet multiple community objectives, particularly historic building reuse and affordable housing.[35] Another is a special variation from the Transfer of Development Rights programs covered in the previous chapter, but directed toward protecting historic buildings and greenfield sites (such as cemeteries, battlefields, and archaeology). For instance, in New York, San Francisco, and Philadelphia, the properties sending/selling development rights for transfer do not have to be open greenspace, but can instead be small-scale city-certified historic buildings. The receiving properties for the development right transfer are still dense high-rise locations in the downtown area. The advantages of a historic preservation-minded Transfer of Development Rights program are to create a financial incentive for reusing smaller, locally certified historic buildings and to shelter the municipality's historic preservation ordinance from court challenges through offering relief to historically-certified property owners.[36] Real estate developers working with new high-density construction in cities with historic preservation-minded Transfer of Development Rights opportunities can also help by purchasing additional development rights from these programs.

Besides financial opportunities and incentives provided by government agencies for reusing historic buildings in community real estate development projects, there are also some provided by non-profit organizations and charitable foundations. Organizations such as the Colorado Historical Foundation and Providence (Rhode Island) Revolving Fund have revolving loan funds that can assist with property acquisition and rehabilitations that can have nominal down payments, fixed interest rates that are negotiable and at or below prime, generous terms and repayment schedules, no early repayment penalties, and collateral options, though a property lien or owner guaranty is often required.[37] Historic preservation non-profit organizations will also sometimes offer consulting services that address community needs related to commercial development on threatened historic properties or affordable housing projects, architectural salvage that assist building reuse projects that incorporate authentically recycled historic materials, as well as serve as receiving partners for the previously mentioned historic preservation easements.[38] A few large non-profit organizations, such as the National Trust for Historic Preservation, the Harden Foundation, and the Sunderland Foundation, have generous grant opportunities for bricks and mortar projects, but these are usually very competitive to acquire and often necessitate a fair amount of project organization prior to proposal submittal.[39] The financial support of many historic preservation non-profit organizations tends to be on education, public outreach, and knowledge development. For local non-profit organizations that do bricks and mortar assistance, such as the New Orleans Preservation Resource Center and Historic Marion (South Carolina) Revitalization Association, the offerings provided are smaller grants for low-income property owners to repair their owner-occupied historic residence, local businesses to restore facades on their historic commercial buildings, as well as tactical urbanism for improving signage, landscaping, lighting, and overall attractiveness, which can be helpful for smaller community real estate development interventions but collectively can have profound impacts.[40]

Lastly, in 1980, the National Trust for Historic Preservation established the Main Street program for purposes of enhancing communities through preservation-based economic redevelopment. The program became incredibly successful with more than 2,000 communities participating. In 2013, the Main Street program became its own independent subsidiary, now called the National Main Street Center. Through locally-based Main Street programs, communities can engage in economic revitalization and real estate development that meets their specific needs and goals through the four point approach that addresses organization, promotion, design, and economic vitality. So, while not a direct form of financial assistance, the Main Street program can be a useful tool for community real estate developers, local businesses, and local government to collaborate together on shared economic revitalization needs that are situated in and next to commercial areas that have a high number of historic buildings.[41]

Human-Centric Approaches

The economic enhancement, physical improvement, and social development of a revitalizing historic neighborhood not only attracts more financial investment but also more affluent people who desire to live there. This is a process defined as gentrification and can be positive when it enhances the quality of life for people in a historic community, especially when economic, racial, age, and gender diversities can be maintained. However, gentrification can become problematic when the influx of wealthier people outprices the ability of original residents to remain, resulting in their displacement that also changes the local culture of place. Because of the historic dynamics between wealth and race in the United States, frequently members of higher economic classes are Euromericans and lower economic classes are people of color, with ancestries from Africa, Latin America, and Asia. In these instances, economic stratification can lead to racial confrontation.[42] Gentrification and the problems of displacement are discussed in greater detail in Chapter 9, but relevant human-centered approaches to historic built environment reuse are relevant here.

In *Human-centered Built Environment Heritage Preservation: Theory and Evidence-Based Practice*, Jeremy C. Wells, the contributing authors, and I explore this very issue and question of what is a human-centric approach to historic building reuse and neighborhood revitalization from a community real estate perspective? There was, of course, a lot that we had to say on the subject, but in summary, it is about obtaining input from residents, stakeholders, and place users when applicable (these are short term customers or tourists) through quantitative and qualitative means in order to develop plans and deliver goods, services, programs, and address current problems according to human needs in a democratic and equitable way. Public participation throughout the development process is essential because situations can change as some projects can take a considerable amount of time to be completed. As our experiences found from the case studies covered within *Human-centered Built Environment Heritage Preservation*, obtaining this kind of information and keeping it updated can be very challenging to say the least and is a practice that few real estate developers engage in or take seriously.[43]

One unconventional human-centered strategy for approaching community real estate development is through the community-built approach, which decentralizes the project among all constituents by empowering people through participation. More specifically, the Community Built Association defines community-built as a "collaboration between professionals and community volunteers to design, organize and create community projects that reshape public spaces. Projects accomplished through community initiative and professional guidance include murals, playgrounds, parks, public gardens, sculptures and historic restorations." Melvin Delgado provides the instructional know-how to do this in *Community Social Work Practice in an Urban Context: The Potential of a Capacity Enhancement Perspective*, where he emphasizes community and urban social work for creating positive community environments in marginalized urban communities.[44] While Delgado and the Community Built Association highlight gardens, murals, playgrounds, and sculpture projects, in *Community-built: Art, Construction, Preservation, and Place*, examples of using the Main Street program for economic development, community-beneficial heritage tourism creation, and historic housing stock renovation, among other ideas are also provided as case studies for more intersectional community real estate development instances that incorporate underutilized historic buildings and neighborhoods. The findings were that when "people are given the opportunity to shape their environment in a way that makes their surroundings more comfortable, more colorful, more personalized, they feel a greater power to make other changes as well,"[45] which is the altruistic objective of community real estate development.

Conclusion

The reuse of historic buildings can be a very effective form of community real estate development, but in order for this to be the case, one needs to know what they have, such as through a historic structures report so that they can adequately understand and plan for their project. Being familiar with local and state historic preservation zoning ordinances and building codes is also important to this process so that legal constraints do not become a burdensome constraint. In order to encourage historic building reuse through proper preservation best practices, the federal government, as well as some state and local governments, have created financial opportunities and benefits for developers and property owners, depending on the objectives of the historic real estate project. Some of these financial opportunities and benefits are historic preservation-specific while others are to encourage community-oriented real estate development that just so happens to take place at old buildings. These various economic incentives for developers can often be coupled together if intelligently strategized. Community real estate development can go beyond traditional property improvement approaches and consider human-centered needs, such as community-built that engages and empowers residents and constituents in the process. Previous examples have found that such approaches can also be very effective at historic properties and thus is an additional method to consider for real estate developers looking to maintain and promote a sense of place that has existed in a community for a considerable period of time.

Notes

1 Joel Shannon, "'We Don't Know If It's Enough': $1 billion May Not Cover Notre Dame Cathedral Rebuilding Costs After Fire," *USA Today*, (20 April 2019), https://www.usatoday.com/story/news/world/2019/04/20/notre-dame-cathedral-fire-1-billion-rebuild-paris-france-church/3528844002/, (1 December 2020).
2 Lucy Davis, "Independence Hall, 1876," *The Encyclopedia of Greater Philadelphia*, (23 March 2019), https://philadelphiaencyclopedia.org/03855v/, (1 December 2020).
3 Edward Rothstein, "Reopening a House That's Still Divided," *The New York Times*, (14 December 2010), https://www.nytimes.com/2010/12/15/arts/design/15museum.html, (29 June 2020).
4 Donovan D. Rypkema, *Feasibility Assessment Manual for Reusing Historic Buildings* (Washington, DC: PlaceEconomics, 2015); Donovan D. Rypkema, *The Economics of Historic Preservation: A Community Leader's Guide*, Third edition (Washington, DC: PlaceEconomics, 2014); Jayne F. Boyle, Stuart M. Ginsberg, Sally G. Oldham, Donovan D. Rypkema, John Leith-Tetrault, and Krista Kendall, *A Guide to Tax-Advantaged Rehabilitation* (Washington, DC: National Trust for Historic Preservation, 2009); Donovan D. Rypkema, *Community Initiated Development: A manual for Nonprofit Real Estate Development in Traditional Commercial Districts*, Second edition (Washington, DC: National Trust for Historic Preservation, 2004); and Donovan D. Rypkema, *Historic Preservation and Affordable Housing: The Missed Connection* (Washington, DC: National Trust for Historic Preservation, 2002).
5 David Listokin, *Landmarks Preservation and the Property Tax: Assessing Landmark Buildings for Real Taxation Purposes* (New York: Taylor & Francis, 2017).
6 Judith Reynolds, *Historic Properties: Preservation and the Valuation Process* (Chicago, IL: Appraisal Institute, 2006).
7 Guido Licciardi and Rana Amirtahmasebi, *The Economics of Uniqueness: Investing in Historic City Cores and Cultural Heritage Assets for Sustainable Development* (Washington, DC: World Bank, 2012).

8 Nur A. R. Demong, Jie Lu, and Farookh K. Hussain, "Personalized Property Investment Risk Analysis Model in the Real Estate Industry," in *Human-Centric Decision-Making Models for Social Sciences*, Peijun Guo and Witold Pedrycz, eds. (Berlin: Springer, 2014), 369–390.

9 Katherine Melcher, Barry L. Stiefel, and Kristin Faurest, *Community-built: Art, Construction, Preservation, and Place* (New York: Routledge, 2017), back cover.

10 Jeremy C. Wells and Barry L. Stiefel, *Human-Centered Built Environment Heritage Preservation: Theory and Evidence-Based Practice* (New York: Routledge, 2019), back cover.

11 Amalia Leifeste and Barry L. Stiefel, *Sustainable Heritage: Merging Environmental Conservation and Historic Preservation* (New York: Routledge, 2018), back cover.

12 Ibid.

13 Witold Rybczynski, *Last Harvest: From Cornfield to New Town* (New York: Scribner, 2014), 43.

14 See Ian Formigle, *The Comprehensive Guide to Commercial Real Estate Investing* (Portland, OR: CrowdStreet, 2017).

15 "Overview of the Real Estate Development Process," *U.S. Department of Housing and Urban Development Archives*, https://archives.huduser.gov/oup/conferences/presentations/hbcu/sanantonio/overview.pdf, (8 December 2020).

16 Mike (Your Property Pro), "The Property Development Process – 8 Critical Steps," *Property Like a Pro*, (2017), https://propertylikeapro.com/property-development-process/, (2 December 2020).

17 "Overview of the Real Estate Development Process."

18 See Deborah Slaton, *Preservation Brief #43: The Preparation and Use of Historic Structure Reports* (Washington, DC: National Park Service, 2005).

19 "Local Preservation Laws," *National Trust for Historic Preservation*, (2020), https://forum.savingplaces.org/learn/fundamentals/preservation-law/local-laws, (2 December 2020).

20 See Thomas C. Jester and Sharon C. Park. *Making Historic Properties Accessible* (Washington, DC: National Park Service, 1993).

21 Stephanie Ryberg-Webster and Kelly L. Kinham, "Historic Preservation and Urban Revitalization in the Twenty-first Century, *Journal of Planning Literature*, 29:2, (2014), 119–39.

22 Paolo Rosato, Anna Alberini, Valentina Zanatta, and Margaretha Breil, "Redeveloping Derlict and Underused Historic City Areas: Evidence from a Survey of Real Estate Developers, *Journal of Environmental Planning and Management*, 53:2, (2010), 257–281.

23 Secretary of the Interior, "The Treatment of Historic Properties," *National Park Service*, (2020), https://www.nps.gov/tps/standards.htm, (3 December 2020).

24 Ibid.

25 "Historic Tax Credits," *Community Developments Fact Sheet*. Washington, DC: Office of the Comptroller of the Currency, (October 2019).

26 "Tax Incentives for Preserving Historic Properties," *National Park Service*, (2020), https://www.nps.gov/tps/tax-incentives.htm, (3 December 2020).

27 "Historic Conservation Easement Program: Frequently Asked Questions," *Restore Oregon*, https://restoreoregon.org/wp-content/uploads/2018/09/Easement-FAQs.pdf, (3 December 2020).

28 "New Markets Tax Credit Program," *U.S. Department of the Treasury*, (2020), https://www.cdfifund.gov/programs-training/Programs/new-markets-tax-credit/Pages/default.aspx, (3 December 2020).

29 "Low-Income Housing Tax Credits," *U.S. Department of Housing and Urban Development*, (2019), https://www.huduser.gov/portal/datasets/lihtc.html, (3 December 2020).

30 "Funding Opportunities," *U.S. Department of Housing and Urban Development*, (2019), https://www.hud.gov/grants/, (3 December 2020).
31 "Tax Incentives," *South Carolina Department of Archives and History*, (2020), https://scdah.sc.gov/historic-preservation/programs/tax-incentives, (4 December 2020).
32 "Mills Act Program," [California] *Office of Historic Preservation*, (2020), https://ohp.parks.ca.gov/?page_id=21412, (4 December 2020).
33 "Ohio Enterprise Zone Program," *Ohio Development Services Agency*, (2020), https://development.ohio.gov/bs/bs_oezp.htm, (4 December 2020).
34 "Tax Increment Financing (TIF)," *Illinois Department of Commerce and Economic Opportunity*, (2020), https://www2.illinois.gov/dceo/ExpandRelocate/Incentives/Pages/TaxIncrementFinancing.aspx, (4 December 2020).
35 "Affordable Housing Streamlined Approval," *San Francisco Planning*, (2018), https://sfplanning.org/sites/default/files/forms/SB35_SupplementalApplication.pdf, (4 December 2020).
36 Donna Ann Harris, "Philadelphia's Preservation Incentive: The Value of the TDR," *National Trust for Historic Preservation*, (October 1992), https://forum.savingplaces.org/viewdocument/philadelphias-preservation-incenti, (4 December 2020).
37 "CHF Revolving Loan Fund: Preservation Loans," *Colorado Historical Foundation*, (2020), https://www.cohf.org/preservation-loans/, (4 December 2020).
38 "About Us," *Providence Revolving Fund*, (2020), https://www.revolvingfund.org/about.php, (4 December 2020).
39 "Special Grant Programs," *National Trust for Historic Preservation*, (2020), https://forum.savingplaces.org/build/funding/grant-seekers/specialprograms?_ga=2.126144077.1902551936.1607094221-561590613.1593797143, (4 December 2020).
40 "Revival Grants," *Preservation Resource Center*, (2020), https://prcno.org/programs/revival-grants/, (4 December 2020; and "Grant Programs," *Historic Marion Revitalization Association*, (2020), https://www.theswampfox.org/facade-grant, (4 December 2020).
41 "Who We Are," *Main Street America*, (2020), https://www.mainstreet.org/about-us, (4 December 2020).
42 A Euromerican is an American with European ancestry.
43 Jeremy C. Wells and Barry L. Stiefel, "Conclusion: A Human-Centered Way Forward," in *Human-centered Built Environment Heritage Preservation: Theory and Evidence-Based Practice* (New York: Routledge, 2019), 317–332.
44 Melvin Delgado, *Community Social Work Practice in an Urban Context: The Potential of a Capacity-Enhancement Perspective* (New York: Oxford University Press, 2000).
45 Kristin Faurest, Barry L. Stiefel, and Katherine Melcher, "Conclusion: Valuing Community-Built," in *Community-Built: Art, Construction, Preservation, and Place*, Katherine Melcher, Barry L. Stiefel, and Kristin Faurest, eds. (New York: Routledge, 2017), 207–216.

6 Community Design and the Real Estate Development Process

Trent Green

What Is Community Design and Why Does It Matter?

Community design is an essential learning module of the Community Real Estate Development Program (CRED) and should be a priority consideration for all involved in the community development process. Communities and neighborhoods in cities across the country are the physical contexts for all types of small-scale real estate interventions that will ultimately help shape a community's future. In established working-class communities, community design often warrants special attention as these areas are more subject to change due to local market conditions and community involvement. The real estate development process is challenged with several connected issues that must be reconciled against the backdrop of pre-existing or established community conditions.

Communities and neighborhoods are the building blocks of a city or town. These residential landscapes comprise a vast majority of their built form and urban fabric. Although all residential areas share several familiar design characteristics, such as formal order, general development patterns and an eclectic visual character, each has a unique sense of identity and place in the city. This is evident in the aggregate conditions encountered in many inner-city and urban core neighborhoods. These contextual conditions define a community's general urban design framework. Depending on locational circumstances, they can have a significant influence on how new development is integrated into existing communities. As community conditions vary from place to place, basic community design characteristics are embedded in their spatial patterns, building architecture and public realm definition. These community design conditions are always present in both urban and suburban neighborhoods.

Single-family neighborhoods are typically characterized by a modular system of rectangular lots with varying street frontage and lot depth dimensions. This standard development pattern allows all structures on a block (except for flag lots) to be directly oriented to the street. This condition is evident in most established neighborhoods. From a community design perspective, the consistent pattern of individual house facades defines the block

DOI: 10.1201/9781003109679-7

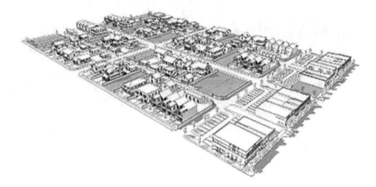

Figure 6.1 Typical Residential Neighborhood Setting.
Image by Mitali Naik.

face of a neighborhood street. When vacant or under-utilized parcels inter-rupt this condition, it often detracts from a neighborhood's general appear-ance (Figure 6.1).

The CRED program introduces students to community design as a basic principle of the real estate development process. This includes an awareness and respect for existing structures, natural conditions and the community values of local residents. This design structure contributes to the percep-tions of place and a strong sense of belonging among local residents.

The physical dimensions of community design encompass a range of con-textual conditions. In low-income and underserved communities, commu-nity design is characterized by a series of integrated physical conditions that are both stable and those that changing. Depending on location, a neighbor-hood's design structure typically includes the following tangible features:

- A spatial order defined by repetitive development patterns of residential blocks
- A modular property subdivision system of individual lots with mostly stand-alone structures
- A discernible rhythm of individual buildings oriented toward the street creating a diverse residential block face
- The general three-dimensional form and heights of structures
- Diverse architectural character and building types with street-facing features (e.g. front porches, gable roofs, main building entries, etc.)
- The general condition of a community's building stock and signs of sta-bility versus blight
- Percentage of occupied structures
- The presence and visual impact of vacant property
- Public realm conditions and how the "street-room" is defined

- The natural setting (e.g. street plantings, front yard foliage, tree canopies, etc.) and the integration of public open space features
- Vehicular and pedestrian infrastructure features and parking coordination
- Prominent neighborhood landmarks and other vertical infrastructure elements

Many of these community design conditions are perceived at different scales, either individually or simultaneously – from individual houses and other buildings on a single parcel, to several structures that define an entire street or block face. The eclectic architectural expressions of a neighborhood's houses and other structures, their general formal composition and the manner in which they engage the street are typically its most discernible community design characteristics. Micro-scale conditions on individual houses, including front yards, porches, roof gables, door and window patterns and general house silhouettes convey a visual narrative of conventional community design. These are the visual conditions that often register with individuals and form the basis for interpretation and impressions (Figure 6.2).

Altogether, these are familiar physical conditions that embody a neighborhood's general character and sense of place. They also define its organizational structure and inform new buildings introduced in existing neighborhoods. This highlights the inextricable relationship between community design and the real estate development process. For existing neighborhoods, their basic order establishes many of the precedent conditions for integrating new infill development. It can also influence real estate project decisions at multiple stages of development, including planning, feasibility analysis, land acquisition and design review. Each stage of the development process will either reference, invoke or prioritize some aspect of community design.

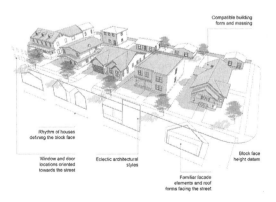

Figure 6.2 Typical Residential Building Block Face with Prominent Architectural Design Features.
Image by Mitali Naik.

Community design is a situational awareness priority for all stakeholders associated with community (re)development and revitalization initiatives. This includes local residents, developers, entrepreneurs, non-profit and market-rate builders and public agencies. Community design is often amplified in low- and moderate-income neighborhoods and has its roots in the 1960s. As older, disadvantaged communities experienced significant physical change through extensive demolition, urban renewal and highway construction, community design also became a resounding mantra for restorative justice and has been closely associated with other societal trends as well. The absence of local residents in the real estate development process and in the formulation of housing policy, coupled with previous discriminatory planning practices and the lack of equitable access to affordable housing, redefined community design as both a cause for change and the result of these actions.

In many instances, these practices would have significant impacts on the extant stability, form and character of low-income neighborhoods across the country – usually in a disproportionate manner. The loss of population, fragmented neighborhood conditions and disinvestment that followed only exacerbated these conditions. This widespread community devastation left many low-income and communities of color in physical stasis and atrophy. For over half a century, some neighborhoods have been challenged by the presence of large swaths of vacant property that has been marginalized by prior institutional actions.

While many of the nation's low-income communities continue to grapple with these issues, some are presented with new market opportunities for a range of (re)development projects. In addition to the scattered-site infill development emerging in these areas, some community-driven initiatives are aimed at restorative and equitable community development. While these efforts are quite varied, they are signaling a greater trend in strategic community development initiatives that yield new single and multi-family residential buildings, as well as other neighborhood-serving commercial uses. Once built, these projects often restore many of a community's lost physical attributes while also addressing other development priorities. Each increment of new infill development provides tangible enhancements along neighborhood streets and blocks and ultimately becomes a permanent community design feature.

Public Policy and the Regulatory Interface with Community Design

Another critical connection in the community real estate development process is its relationship with public policy and its interface with development regulations. Community revitalization is promoted at multiple levels by local governments through comprehensive plans, land-use policies, strategic planning initiatives, and building regulations. Various aspects of community design are integrated into each of these tools. At the policy level, community design standards are typically incorporated in a jurisdiction's comprehensive or general plan. Its housing and land-use elements identify goals and objectives for housing, community development and a range of other issues over a long-term horizon. Relative to community development, the comprehensive plan is the foundation that guides all zoning and land-use decisions. Its policies outline protections for existing neighborhood conditions and the preservation of sound housing stock, as well as goals for attainable and elderly housing and performance standards for infill development. Long-range plans, zoning and land-development regulations and historic district guidelines all emanate from this document. All of these policies have a direct influence on the form and character of existing neighborhoods.

In addition to the comprehensive plan, the most familiar tools that directly influence community real estate development are zoning and land-development regulations. These regulations establish minimum compatibility criteria for residential densities and building types, standards for site development and other requirements that directly affect community design conditions. In established neighborhoods, this equates to infill buildings achieving a type of "diagrammatic compatibility" with adjacent structures along a block face. From an architectural perspective, zoning does not imply design or stylistic preferences, nor does it infer design quality. However, in historic districts, planning overlay districts and areas where master plans are in use, design standards for new development may also be regulated to protect or enhance existing community design conditions, or to realize desired outcomes.

Other regulatory tools being adopted by local jurisdictions to assist in the community development process are what's commonly referred to as form-based codes. These are regulatory tools that *"… foster predictable built results and a high-quality public realm by using physical form (rather than separation of uses) as the organizing principle for the code" (Form-Based Codes Institute).* Form-based codes may be used in conjunction with or replace conventional zoning altogether. They incorporate design standards for new development that are derived from the physical conditions of the host neighborhood, such as general development patterns and formal building properties as a framework for defining infill compatibility. Form-based codes are graphic-based regulatory tools that incorporate simple diagrams as a flexible way to clarify preferred future conditions. These tools operate at different scales of a community, including the neighborhood level, the street or block face and at the level of individual parcels. They frequently incorporate qualitative community design criteria for new infill projects, as well as other accepted sustainable development practices. Where used, form-based codes apply to both market-rate and affordable housing projects, and to neighborhood commercial and mixed-use projects. Over the past four decades, form-based codes have been strategic in helping preserve community attributes, pairing voids in neighborhood fabric and ensuring new development achieves a good community design fit. Their focus on design allows all stakeholders in the community real estate process to be equally knowledgeable about the benefits and the impacts of new projects early in the development process. Combined with other innovative practices, form-based code usage is widely considered a smart and sustainable approach to community development.

Community plans are gaining popularity in the community real estate development process. These are tools that are often adopted to empower local residents to defining their own aspirations and visions for neighborhood preservation, design and long-term growth. Community plans are not regulatory in nature but have direct implications on spatial conditions in both urban and suburban enclaves. They often serve as a basis for neighborhood preferences in resolving contested land use and other zoning decisions.

Over the last several decades, several jurisdictions have removed many of the zoning and other restrictions that have posed impediments to creating affordable, attainable and workforce housing in existing neighborhoods. These regulatory obstacles have been identified in local redevelopment initiatives, as well as the U.S. Department of Housing and Urban Development's (HUD) 2021 policy *"Eliminating Regulatory Barriers to Affordable Housing: Federal, State, Local, and Tribal Opportunities"*. They include federal, state and local recommendations and incentives for coordinating consistent and streamlined review processes, re-using existing structures and other measures to reduce construction costs. In these instances, real estate development and the ongoing shortages of housing supply are directly related to community design. Many of the measures now in use allow for greater densities and a broader range of residential building types deemed appropriate in existing neighborhood contexts. They also allow for more creative solutions to infill development and a broader fluence on community design over a longer period.

Community Design and New Housing in Existing Neighborhoods

Infill development essentially re-purposes vacant, obsolete and under-utilized property in established neighborhoods with new construction. It is a widely utilized approach for crafting complete and vibrant neighborhoods and expanding the supply of housing at different price points. This process reflects the successive increments of new development that natural to community growth and revitalization. In many low-income and disadvantaged communities, infill development is characterized by either small-scale residential buildings on scattered sites or an intensification of residential densities to maximize the development capacity of available property. In addition to these, infill development has been initiated to stabilize existing neighborhood conditions, bolstering local economic development, enhance public health, maximize the use of existing infrastructure and to create densities that support local public transportation. Regardless of motivation, all infill development has a direct effect on the form and character of an existing neighborhood.

The practice of redeveloping existing neighborhood structures and fill-ing in vacant lots with new structures is an ongoing process. This process can directly influence community design in both a subtle and more noticea-ble manner. For many small-scale infill projects, the basic approach hasn't changed much over the last half a century. In many communities, individual "missing teeth parcels" along a block face are acquired and improved with a primary or secondary residence (e.g., rear yard cottages, garage apartments, etc.). Other small-scale infill projects may involve the assemblage of adjacent parcels to construct multiple houses, a multi-family apartment building, a stand-alone commercial structure or a neighborhood mixed-use structure. Infill parcels can vary in size and configuration from one neighborhood to another; however, they do have one consistent characteristic. In most com-munities across the United States, the standard single-family house lot de-fines a standard increment for new development. Community builders focus on these and other under-utilized properties as opportunities to implement projects that often stitch together or repair a neighborhood's urban fabric. In established neighborhoods, these projects are often inspired by surrounding buildings. Many employ skillful architectural design techniques to achieve an acceptable degree of compatibility, make unique visual statements or to better blend in with the immediate context.

On the surface, this type of development seems like a simple process. However, it can be complicated by several factors, most notably neighbor-hood acceptance and who actually benefits from it. In addition to all the stated benefits associated with new housing in existing disadvantaged neigh-borhoods, infill also invites new residents. Although each project has its own motivation, certain types of infill development are often singled out as a leading contributor to rising property taxes and the displacement of low-income residents. These and other social dynamics associated with some in-fill development may hasten significant changes in neighborhood form and character.

Although all communities have some physical attributes, low-income ar-eas have often endured decades of disinvestment, depressed property values and a negative image. After decades of being passed over, many of these areas are being re-assessed for new investment opportunities because of changing market perceptions, availability of reasonably priced land and proximity to other community assets. To redevelop these properties with new housing options and help stabilize these areas, new strategies and part-nerships are being formulated between non-profit community builders, pub-lic agencies and private investors. One of the major distinctions of these partnerships is their inclusion of "legacy" residents – giving some projects a more community-driven focus (Figure 6.3).

To maximize housing options in existing neighborhoods and foster a more diverse demographic makeup, low-density infill strategies today may include more than just single-family structures. They are now likely to in-corporate additional residential building types that define the middle range of the housing spectrum, such as duplexes, four-plexes, garden apartments,

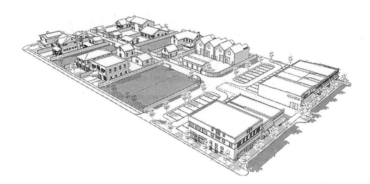

Figure 6.3 Incomplete Neighborhood Fabric Characterized By "Missing Teeth" and Under-utilized Property.
Image by Mitali Naik.

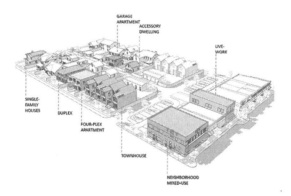

Figure 6.4 Single-Family and Multi-Family "Missing Middle" – Low-Density Infill Residential Building Types.
Image by Mitali Naik.

townhouses and small apartment buildings. Although most communities comprise single-family residences, until recently many neighborhood scale, multi-family building types have not been a part of the general community fabric. In fact, many jurisdictions are actively encouraging the use of these building types to promote equitable access to new housing for legacy residents, maximize the use of existing infrastructure and to realize densities that support walkability and public transit (Figure 6.4).

Missing middle residential building types are intended to address the gaps in local housing markets and fit seamlessly in existing detached single-family neighborhoods. When used in combination throughout a neighborhood,

Figure 6.5 Complete Neighborhood Fabric with Existing and New Infill Buildings. Image by Mitali Naik.

these multi-family building types provide a flexible method for increasing densities without adversely affecting its design structure (Figure 6.5).

The Impacts of Infill on Community Design

New development can have a dramatic impact on the general design character and form of physical conditions in a community. This is often more noticeable in low-income and disadvantaged communities that have experienced generations of disinvestment and erosion of building stock. With the increasing demand for both owner-occupied and rental housing options and the influence of local real estate markets, many of these communities are becoming more appealing to individual investors, community builders and for-profit developers. Recognizing the multiple benefits of associated with infill development, public sector agencies are also taking a more active role in these initiatives by devising long-term revitalization plans, providing regulatory incentives and in many instances real property. While scattered-site infill development is part of the natural cycle of community evolution, many of the coordinated efforts around the country are proving to be neighborhood catalysts and agents for transformative change. In these instances, new structures are better calibrated with the general design structure and character of the host neighborhood. In addition to this, the resulting higher densities from these projects are expanding opportunities for equitable access to new housing by long-term residents. This is increasingly evident in infill projects completed by community builders and non-profit entities across the county. Many of these projects have a major community design presence and are demonstrating a commitment to "community-driven" development. This type of development is often positioned to realize a balance between housing production, community and social values.

This is evident in the real estate practices of Develop Detroit, a six-year-old non-profit entity established to provide citizen inspired redevelopment in many of the city's distressed neighborhoods. Detroit is well known as a major American city that has suffered from massive loss of population, disinvestment, neighborhood abandonment and deterioration. Since inception, Develop Detroit has implemented multiple infill projects in neighborhoods that have not seen any (public nor private) investment in decades and characterized by extensive vacant lots and building abandonment. Using a tri-sector (public, private and foundation) partnership model, their work in Detroit's North End neighborhood takes a holistic and scaled-up approach to community development to reduce blight and property vacancy.

Since 2015, this organization has combined its private real estate development activities with a social agenda to create mixed-income housing for

Figure 6.6 North End Neighborhood – Previous Conditions.
Courtesy of Christian Hurttienne Architects.

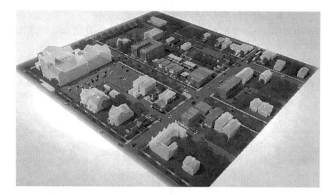

Figure 6.7 Proposed Infill Build-Out.
Courtesy of Christian Hurttienne Architects.

Figure 6.8 Single-Family Prototype Design.
Courtesy of Christian Hurttienne Architects.

Figure 6.9 Multi-family Prototype Design.
Courtesy of Christian Hurttienne Architects.

legacy residents and other prospective homeowners throughout the city. Their work has helped stabilize several neighborhoods and helped bring economic development to parts of the city devastated by intrinsic societal and market trends. The North End Homes project focuses on many of the area's vacant parcels and abandoned buildings within a four-block area. Develop Detroit's strategy in his instance focused on a hybrid approach to community development that will help re-invigorate this area. This project was intended to bring new owner-occupied single-family structures, affordable multi-family structures, renovated houses and neighborhood mixed-use buildings to the local market. By clustering their efforts, this project is having a transformative effect on the immediate neighborhood and is contributing to its diversity by providing a range of housing options at different price points for prospective homeowners and renters.

This project exemplifies the impacts that new development can have on existing neighborhoods. Not only does it introduce new housing in an otherwise stagnant market but it also provides the added benefit of restoring the neighborhood fabric. While most of the new units in this project were constructed in multiple phases at market-rate prices for part of the city, each phase of development incorporates an affordable or below market-rate component (Figures 6.6–6.9).

7 Community Redevelopment, Tax Increment Financing and CRED

Jeff Burton

Community Redevelopment

Community redevelopment is the reinvention of cities through the development and utilization of urban land, change of use over time, demolition, and repurpose (Hersh, 2018). This activity is also the state-empowered, local act of eliminating and preventing the development or spread of slums and blight (Ball & Maginn, 2005). Community redevelopment is a century-old, national public policy designed to eliminate or prevent urban slums and blight (Gordon, 2003) and occurs worldwide (Hersh, 2018; Martin, 2018). Local government leaders have struggled for over a century to define blight and apply effective measures to mitigate and negate its community impact. The term includes vacant lots, abandoned or dilapidated structures, and brownfields (Schilling & Pinzón, 2016). Community redevelopment activities have been decidedly economic, trading off environmental and social concerns (Lefcoe, 2011). However, contemporary community redevelopment evolves an inclusive affinity toward sustainability's social, economic, and ecological systems (Križnik, 2018).

The American Planning Association (2004) notes that community redevelopment occurs when private interest to improve a deficient area falls short and is stimulated by public action. Community redevelopment commonly occurs through a designated redevelopment authority, using tax increment financing, or TIF, in a designated place, and according to a locally adopted redevelopment plan (Cook, 1959; Jonas & McCarthy, 2009). Roberts and Sykes (1999) add that community redevelopment is inevitable as economic, political, and social networks generated new economic and civic growth demands. In their view, community redevelopment is also beneficial because it creates opportunities to address the area's conditions.

State Authority

Since the Nixon administration, The Federal Government has abstained from local community redevelopment. At the beginning of Nixon's second term in January of 1973, HUD Secretary George W. Romney announced a moratorium on all federal housing funds, and during a presidential speech

DOI: 10.1201/9781003109679-8

in September of the same year, Nixon categorically unfunded future urban renewal and threatened a veto on any new congressional spending bills programs other than defense and budget deficit reduction (Morris, 1974).

After the federal housing act's demise and diminished national oversight, most states passed some individualized urban renewal programs. The states did not boldly go where the federal law had gone before but clung to existing "blight" and "but for" strategies (Weber & O'Neill-Kohl, 2013). Sticking to this path of least resistance allowed local government to refine the legality separating public funds from private uses (Rast, 2012). Forty-nine out of 50 states still operate varying community redevelopment types, coalescing without national guidance (Leigh, 2003). Some states completely rewrote their existing laws with newer versions, while other states continue to amend laws enacted almost a century passed. For example, in 1977, the State of Illinois updated its urban renewal and TIF laws as local governments reacted to federal support erosion (Reingold, 2001). Weber and O'Neill-Kohl (2013) identified federal redevelopment programs' transfer to state control on urban renewal administration professionals' vested interests. These consultants branded program markers from the past onto new state legislation. They enlisted new supporters, making it challenging to distinguish original federal urban renewal practices from the more unique entrepreneurial policies that exist today.

Today's Blight

Today, community redevelopment experiments with new concepts such as New Market Tax Credits, Opportunity Zones, and synthetic TIF. They build on previous developer experiences and could serve to help future redevelopers (Cabanella, 2018). A new blight definition identifies a state in which structures deteriorate and deteriorate, and economic distress or the endangerment to life or property (Hipler, 2007). Moore (2005) defines the opposite of blight as a "healthy job-rich environment, with a greatly increased tax base." Rarely given an adequate definition, local governments have been allowed flexibility in their interpretation and determination (Gordon, 2003). Blight is the prime catalyst for both brownfield remediation and community redevelopment (Alker, Joy, Roberts, & Smith, 2000).

The Redevelopment Area

The Federal Housing Act of 1949 initially required that redevelopment sites be designated as a slum, blighted, or deteriorated. The area was also selected according to the city's general plan. Once established, this area would be purchased or condemned by a local public body after a public hearing. (81st Congress 1st, 1949). At community redevelopment's inception, blight was considered as "the grotesque living conditions or deteriorated industrial areas" in urban communities across the nation (Association, 2004).

Modern redevelopment areas, sometimes called a tax increment district, or TID, establish the geographic area from which redevelopment activities occur. Johnson and Kriz (2019) note that the boundaries should include enough property to generate the TIF needed to pay for the projects. They add that only parcels that receive the redevelopment benefits should fall within the district. Modern blighted areas have "underperforming or obsolete land uses or buildings that exist where the rational private investment cannot overcome market forces. In such situations, disinvestment and blight may have already occurred or are deemed to occur" (Association, 2004). The APA adds that a publicly defined delineation process must be in place before creating the redevelopment area. This process should exist for slum and blight removal and eliminations and public-private partnerships.

The Redevelopment Authority

According to Roberts and Sykes (1999), the local authority is vital in urban community redevelopment. This role may include the landowner, the planning authority, or both. The local governing body establishes the redevelopment authority and structure. This structure provides administrative, legal, financial, and organizational support for redevelopment activities. The redevelopment board is usually appointed by the local elected officials and governs the authority. The authority may enter into contracts and financial debt to carry out redevelopment (Johnson & Kriz, 2019).

The Redevelopment Plan

According to Roberts and Sykes (1999), planning systems are very mechanical, yet planning is also a powerful tool for sustainable redevelopment. They continue that, too often, local plans complete with out-of-date land-use zoning ambitions, urban density, car parking, and transportation standards. They argue that community engagement is a fundamental social precondition to redevelopment planning. The local planning process should be framed as a holistic community process emerging from a wide-ranging discussion of neighborhood needs and wants before the drafting.

The federal government initially required local governments to create a redevelopment plan before slum clearance and private sector resale or lease or public projects development (81st Congress 1st, 1949). In most states, the redevelopment plan is prepared by the redevelopment authority and ratified by its governing body (Hill, 1952).

The APA recommends that cities not perform redevelopment practices if they do not have a comprehensive plan. They also suggest that redevelopment plans conform to such a plan (Association, 2004). The redevelopment plan should contain all of the information needed by a stakeholder, especially taxpayers and their elected representatives, to support the redevelopment actions rationally. Schacht (2018) states that the plan typically includes the entire assessment process, planning, design, and implementation. He adds that this effort engages necessary infrastructure improvements and creates the functional urban form to create new human activities. This information includes project and program details and a description of the area after redevelopment. Estimated property values and annual taxes should be included (Johnson & Kriz, 2019).

Turok and Shutt (1994) identify a dysfunction in urban problem-solving. They argue that local planning efforts address redevelopment issues in a piecemeal manner and do not consider the linkages between the economy, environment, and society. Countering this criticism, Alden and Boland (1996) promote spatial planning. They argue that this methodology expands the purely physical redevelopment and land-use planning to include sustainable development systems. Roberts and Sykes (1999) argue that a strategic management system should be created to identify the area's sustainable development goals. These goals make a framework for authorities to derive their redevelopment plans. They also note that the area's planning system influences physical problems in urban areas. They report that blight and neglect have resulted from over-ambitious, enabling planning systems.

Community Redevelopment Criticism

der Krabben and Needham (2008) note the difficulty and lengthy time needed to process redevelopment. They identify the responsibility of three factors that influence the time required to assemble properties and develop the necessary redevelopment capital. First, the land and its ownership are usually fragmented or it is held for speculative opportunities. Second, community redevelopment usually includes expensive public infrastructure and services such as parks, roads, water, and sewer.

These assets may increase the land values in their area and are difficult to assess. Third, the redevelopment area may be challenging to outline as areas surrounding the site may have a similar disposition.

Jonas and McCarthy (2009) identify five drawbacks to community redevelopment. First, community redevelopment projects may not create the needed revenues to pay the debt service. Second, community redevelopment draws future tax revenues from other public services like schools and hospitals. Johnson and Kriz (2019) disagree, stating "this may have been true at the beginning of TIF, many places have learned how to protect" existing public services. Third, when community redevelopment focuses on public/ private partners, the authority may neglect the community's social needs, such as affordable housing. Fourth, unnecessary use of community redevelopment funds cuts into its governing body's general reserves. Fifth, due to the lack of voter approval, community redevelopment borrowing may lack transparency.

Redevelopment authorities have also been criticized for banking TIF and establishing political power, acting like semi-autonomous governments, and allocating resources to favored projects (Johnson & Kriz, 2019). The authors continue that standard finance practices should be followed to evaluate the economic feasibility of redevelopment practices, especially property values.

The APA identifies four current trends that affect community redevelopment. First is state legislative action and case law threatening the local government's ability to implement community redevelopment. This trend has weakened local agencies' power, strengthened citizen participation, expanded legal challenge opportunities, and altered TIF revenue sharing strategies. Second, the Association reports that there is significant administrative and policy inconsistency between the states. Third, it appears that there is an increased number of local government officials that view community redevelopment as more of an economic tool rather than a community revitalization catalyst. They found many cases that local governments were chasing sales and property tax revenues. Fourth, the organization and failing "to recognize the role of the redevelopment as a tool for creating a sense of place on a community's unique natural and cultural assets." acknowledges a diminishing relationship between community redevelopment and comprehensive planning (Association, 2004).

According to Cabanella (2018), gentrification occurs when "a neighborhood becomes so successful its own residents are among the less affluent who can no longer afford to live there." Community redevelopment is consistently under critical fire for gentrification and resident displacement (Johnson-Ferdinand, 2015; Kim, Marcouiller, & Choi, 2019) and farmland targeting for redevelopment (Blount, Ip, Nakano, & Ng, 2014). Hersh (2018) agrees that today's most significant community redevelopment issue is gentrification. He describes gentrification as a "neighborhood becomes so successful its residents are among the less affluent who can no longer afford to live there." With gentrification comes the loss of community cohesion, forced displacement, and lack of low-income housing (Kim et al., 2019). Cabanella (2018) argues that the lack of community redevelopment attention to an underserved community leads to the thought that "the only thing worse than gentrification, in no gentrification." She adds that for positive community redevelopment to occur, the community must be genuinely engaged. The decision-making process must listen to the area's residential voice regarding progress and preservation.

Tax Increment Finance

TIF has become the most widely used local government economic development financing tool and redevelopment practice in the United States (Briffault, 2010; Johnson & Kriz, 2019; Paull, 2008). Public finance economists classify the method as a prime example of "value capture" (Youngman, 2011). Generally, though enduring many changes over its lifespan, TIF is still the taxable revenue of planned economic growth from (re)development within a geographic area, sometimes called a tax increment district (TID). TIF is set aside for a set period to finance physical infrastructure improvements and other expenditures intended to spur the economy.

The TIF theory assumes that tax growth generated within a designated area or tax increment district (TID) remains there for some time and is expended for physical "bricks and mortar," infrastructure, and other expenses generated economic development. Though highly controversial, TIF is seen by many as a form of "self-funding" created without raising new property taxes. TIF can be used as a "pay-as-you-go" financial solution or debt service (Johnson & Kriz, 2019). Most TIF funding is derived from real property values, and approximately 18 states also use non-ad Valorem taxes, such as sales tax, to fund TIF projects (Briffault, 2010).

The Structure

The monetary value of all real property in the TID is valued as a fixed date known as the "frozen value." Taxing authorities, contributing to the increment, continue to receive frozen value property tax revenues. These frozen tax revenues are for general government purposes. However, any real property value revenues increase in the TID referred to as "increment" are deposited into the TID and dedicated to its plan.

The Area

Community redevelopment authorities usually operate a TID with a redevelopment commission providing oversight and governance. These areas typically overlay other taxing authorities that provide their funding. Briffault (2010) and Diamond (1983) point out that in two ways, TIDs have the same ideology as 19th century "special assessment districts." First, both districts are created based on local improvement impact. Second, those public improvements are specific to the areas of sourced funding. One primary difference between the two districts is that TIDs collect revenue from existing tax millage, whereas special improvement districts use an additional millage. A second difference is that special assessment districts are used for public infrastructures such as sewers, water, and streets. TID revenue may be used for private enterprise and profit-driven projects (Weber & O'Neill-Kohl, 2013).

Public Use

State legislation must conform with its constitution, which requires that taxing districts spend public tax funds for public uses (Schoettle, 2003). The public purpose is rarely legislatively defined, although legal cases have given the term more definition. Weber and O'Neill-Kohl (2013) note that this ambiguity allows local government advantage providing questionable public use schemed flexible private sector project financing. They add that the "public use" definition extension came from state urban renewal laws written to match federal urban renewal laws. Two TIF local spending tests have been used to legitimizing public purpose spending. First, the area may be pre-tested for slums and blight. A sanitization process to eradicate the stigma and take the area back to its "empty state" was diagnosed to guarantee that blight would not return (Weber, 2002). Second, a "but for" test may be applied to projects to demonstrate that "but for" the TIF, the work would not occur. Tracking this test, though challenging, is found in some state laws. The statutes require the locality to resolve that the TID would not redevelop to its highest and best use without TIF assistance (Weber & O'Neill-Kohl, 2013). TIF is an essential part of many local governments' financial, political, and administrative structures. TIF offers local governments a funding source for a myriad of uses.

TIF Criticisms

TIF benefits a blighted area by focusing the property or sales tax revenue generated in the area to remediate brownfields, acquire land, improve infrastructure, and incentivize business location (Greenbaum & Landers, 2014). Supporters argue that TIF revenues are "self-funding" and benefit their sourced areas (Parker, 2012). Some say that properties in proximity to stabilizing TIF areas see an increase in value (Yadavalli & Landers, 2017). Greenbaum and Landers (2014) concur, noting numerous studies that have found a "positive association" between TIF areas and property values (Figure 7.1).

Lefcoe (2011) notes that TIF enables the raising of capital by local governments to "jump-start" private sector investment and their tax base. He identifies TIF as a "win-win-win" scenario for municipalities, developers, and taxpayers. City leaders may claim responsibility for new private capital investments, the private sector gains public investment in improving their project, and tax roll enhancement and job creation benefit the public.

Numerous states have enabled TIF abuse when they do not require a blight finding or do not create a quantifiable definition. There is a growing trend to remove the verbiage and replace it with a general determination of potential economic development (Johnson & Kriz, 2019).

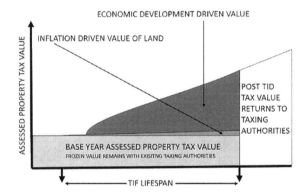

Figure 7.1 Tax Increment Finance Structure.

The majority of TIF-relevant literation focuses on the "but for" question (Yadavalli & Landers, 2017). For example, research in Chicago finds that most TIF districts fail to meet the "but for" test (Lester, 2014). Greenbaum and Landers (2014) argue that the "but for" test is an insufficient economic trend measurement and that cost/benefit growth measurement should be used to evaluate TIF use. Yadavalli and Landers (2017) note that TIF critics argue that economic development would have occurred naturally without the TIF and the "but for" determination.

Brooks, Plante, and Hayes (2017) also argue that private sector investment using public monies is hugely controversial. That argument trickles down into more specific criticisms, including the diversion of general tax funds from schools and local governments (Joravsky & Dumke, 2015; Lester, 2014), income and race discrimination (Joravsky & Dumke, 2015), public sector giveaways (Brooks et al., 2017) induced sprawl, and corporate blackmail (Good Jobs, 2019). During the City of Chicago's Mayor Emanuel administration, 48% of TIF investment was committed to about 5% of its geographic area and 11% of its population (Joravsky & Dumke, 2015).

Once a method to redeem blighted urban areas, critics now argue that TIF designations have adapted for economic development in non-blighted areas (Chapman & Gorina, 2012; Kerth & Baxandall, 2011). Many states have diluted the blight prerequisite to the point of irrelevance (Johnson & Kriz, 2019). Conversely, Lefcoe (2011) argues that TIF restriction to blight inhibits overall economic growth. Chapman and Gorina (2012) also identify a lack of transparency, accountability, and temporality within TIF activities.

Johnson and Kriz (2019) identified several areas in which TIF has not lived up to expectations. First, the incremental ad valorem value may not meet growth expectations and not produce necessary debt service. Second, the base value may not provide adequate local government funding and not address annual average growth. Third, some states do not sunset TIF districts with an effective termination date.

Community redevelopment TIF may fund brownfield redevelopment projects (Haslam, 2009). Florida, Tennessee, and Michigan authorize TIF brownfield remediation (USEPA, 2014). Only Florida allows for TIF incentivized brownfield redevelopment (Florida, 2018).

Community Real Estate Development (CRED)

Community Real Estate Development (CRED) is capital-intensive and multidisciplinary. This process produces risk and requires an entrepreneurial attitude. CRED housing includes rental made up of large and small complexes, single-family, duplexes, affordable, workforce, particular need, and senior housing. For purchase, shelter and transitional housing are also inclusive to the process. Commercially, food security, charter schools, health centers, human service and training facilities, business incubators, community amenities, and shopping areas are also allowable CRED uses (Johnson, 2020).

The National Scope of Community Redevelopment and TIF

Many of these proposed "community" real estate developments are presented for existing underserved urban areas, usually identified as "blighted" or "slum" areas by the state in which they reside. Forty-nine out of the 50 United States maintain some form of slum or blighted area redevelopment legislation. Only California has a legal vacuum in this area. This state once supported the most lucrative, sophisticated community redevelopment protocols in the county. The state disbanded its "RDAs" in 2012 due to state financial limitations. Of those 49 laws, 42 of them mention slums, and all of them cite blight. Thirty-five of the statutes reference tax increment financing.

Before selecting a proposed site, the developer must discover if the state empowers community redevelopment, TIF, and other capital opportunities. This discovery is achieved through a review of these state legislation. These laws typically define terms such as slum, blight, and redevelopment. These legislations also determine area designation, authority board makeup and powers, and spending stipulations. Within this dialogue, the developer may determine locations and housing, commercial, and infrastructure financing opportunities.

Communicating to Government and Politics

As was mentioned earlier, with their historical roots embedded in past federal housing and urban renewal laws, most contemporary redevelopment authority plans should be compliant with their governing body's comprehensive plan. The Community Real Estate Developer should respect this synergy and review this larger plan also. The more significant the jurisdiction measures the real estate development, the more criticism it may receive; it is vital for a critically reviewed product to address the redevelopment area's needs and general jurisdiction. This professional developer process assists the plan through planning, zoning, and the political processes and aligns the proforma with available incentives.

Types of Incentives

Each state empowers some entrepreneurial power level to its local county and municipal bodies, while other states manage the development process directly. Some of these powers are state-wide and include tax credits, sales tax rebates, first-time homebuyer assistance, loan guarantees, to name a few. Some of the incentives are derived locally and are specific to that city or county. Residential opportunities are usually tied to an affordability index, usually as a percentage of the local average median income (AMI). Commercial benefit, many times, is measured to future TIF (real property value improvement) or job creation. Sophisticated community redevelopment authorities are now aligning local incentives to some form of triple-net (economic, environmental, social) benefit nexus, incentivizing developers to create sustainable developments and not just maximizing profit.

Present Value of Future Incentive Value

These financial encouragements should address CRED gap financing and should be proforma inclusive. If provided as a grant, the incentive funds should be valued as owner's equity in the pro forma and capital stack (gap financing) or as a loan if the payback is required. When considering how to apply these values, the terms are the key. Are the incentives reducing government application fees, such as impact fees, permit fees, or service fees, or are they cash payments? Do the incentives come after the project's certificate of occupancy (C.O.)? Are they paid over time or all at once? If the stimuli are delivered over a term of years, the proforma should use adjust to number to present value (P.V.). The P.V. is the calculated current value of a future amount of money or cash flow stream, given the return rate (RoR). Future cash flows are discounted at the rate, and the higher the discount, the lower the P.V. Some states have cottage financial lenders developed just for these types of incentives.

Works Cited

81st Congress 1st, S. (1949). *Summary of provisions of the National Housing Act of 1949.* Retrieved from Washington D.C.: https://web.archive.org/web/20160215080101/, https://bulk.resource.org/gao.gov/81-171/00002FD7.pdf

Alden, J., & Boland, P. (1996). *Regional development strategies: a European perspective* (Vol. 15): Psychology Press.

Alker, S., Joy, V., Roberts, P., & Smith, N. (2000). The definition of brownfield. In (pp. 49). Great Britain: CARFAX PUBLISHING CO.

Association, A. P. (2004). APA policy guide on public redevelopment. Retrieved from https://www.planning.org/policy/guides/adopted/redevelopment.htm

Ball, M., & Maginn, P. J. (2005). Urban change and conflict: evaluating the role of partnerships in urban regeneration in the U.K. *Housing Studies, 20*(1), 9–28. doi: 10.1080/0267303042000308705

Blount, C., Ip, W., Nakano, I., & Ng, E. (2014). *Redevelopment agencies in California: history, benefits, excesses, and closure/Casey Blount, Wendy Ip, Ikuo Nakano, Elaine Ng.*

Briffault, R. (2010). The most popular tool: tax increment financing and the political economy of local government. *University of Chicago Law Review*, 65. Retrieved from http://ezproxy.lib.usf.edu/login?url=http://search.ebscohost.com/login.aspx?direct=true&db=e dshol&AN=edshol.hein.journals.uclr77.7&site=eds-live

Brooks, R., Plante, C., & Hayes, D. W. (2017). Something for nothing? Understanding tax increment financing. In (p. 48). Association of Government Accountants.

Brownfield Redevelopment Act, § 376.77–376.85 (2018).

Cabanella, G. L. (2018). Revitalizing neighborhoods, housing and social equity. In B. Hersh (Ed.), *Urban Redevelopment: A North American Reader* (pp. 113–124). New York: Routledge.

Chapman, J. I., & Gorina, E. (2012). Municipal fiscal stress and the use of Tax Increment Financing (TIF). *The Town Planning Review, 83*(2), 195. doi:10.3828/tpr.2012.11

Cook, J. F. (1959). The battle against blight. *Marquette Law Review, 43*, 444.

der Krabben, E. v., & Needham, B. (2008). Land readjustment for value capturing: a new planning tool for urban redevelopment. *Town Planning Review, 79*(6), 651–673.

Diamond, S. (1983). The death and transfiguration of benefit taxation: special assessments in nineteenth-century America. *The Journal of Legal Studies, 12*(2), 201–240.

Good Jobs, F. (2019). [Sears and "Job Blackmail:" an Early Warning Sign of Job Loss?]. Web Page.

Gordon, C. (2003). Blighting the way: urban renewal, economic development, and the elusive definition of blight. *Fordham Urban Law Journal*, 305. Retrieved from http://ezproxy.lib.usf.edu/login?url=http://search.ebscohost.com/login.aspx?direct=true&db=e dshol&AN=edshol.hein.journals.frdurb31.24&site=eds-live

Greenbaum, R. T., & Landers, J. (2014). The tiff over TIF: a review of the literature examining the effectiveness of the tax increment financing. *National Tax Journal, 67*(3), 655.

Haslam, C. (2009). Urban redevelopment and contaminated land: lessons from Florida's brownfield redevelopment program. In (pp. 153). Great Britain: Cambridge University Press.

Hersh, B. (2018). Chapter 1 history and trends. In B. Hersh (Ed.), *Urban Redevelopment: A North American Reader* (First edition. ed., pp. 1–20). New York: Taylor & Francis.

Hill, P. H. (1952). Recent slum clearance and urban redevelopment laws. In (Vol. 9, pp. 173–189).

Hipler, H. M. (2007). Economic redevelopment of small-city downtowns: options and consideration for the practitioner. *The Florida Bar Journal, 81*, 39.

Johnson, C. L., & Kriz, K. A. (2019). *Tax increment financing and economic development: uses, structures, and impact*: SUNY Press.

Johnson, E. (2020). Introduction to community real estate development. In Tampa, FL: University of South Florida, Florida Institute of Government.

Johnson-Ferdinand, A. V. (2015). *Spatial decision support systems for sustainable urban redevelopment*. Retrieved from http://pqdt.calis.edu.cn/detail.aspx?id=Wk1PXkpEAvc%3d (Dissertation/Thesis)

Jonas, A. E., & McCarthy, L. (2009). Urban management and regeneration in the United States: state intervention or redevelopment at all costs? *Local Government Studies, 35*(3), 299–314.

Joravsky, B., & Dumke, M. (2015). Who wins and loses in Rahm's TIF game. *Chicago Reader*.

Kerth, R., & Baxandall, P. (2011). *Tax-increment financing: the need for increased transparency and accountability in local economic development subsidies*. US PIRG Education Fund.

Kim, H., Marcouiller, D. W., & Choi, Y. (2019). Urban redevelopment with justice implications: the role of social justice and social capital in residential relocation decisions. *Urban Affairs Review, 55*(1), 288–320. https://uar.sagepub.com/content/by/year

Križnik, B. (2018). Transformation of deprived urban areas and social sustainability: a comparative study of urban regeneration and urban redevelopment in Barcelona and Seoul. In (Vol. 29, pp. 83–92).

Lefcoe, G. (2011). Competing for the next hundred million Americans: the uses and abuses of tax increment financing. *The Urban Lawyer, 43*(2), 427. Retrieved from http://ezproxy.lib.usf.edu/login?url=http://search.ebscohost.com/login.aspx?direct=true&db=e dsjsr&AN=edsjsr.41307752&site=eds-live

Leigh, N. G. (2003). *The state role in urban land redevelopment*: Brookings Institution Center on Urban and Metropolitan Policy.

Lester, T. W. (2014). Does Chicago's tax increment financing (TIF) programme pass the 'but-for' test? Job creation and economic development impacts using time-series data. *Urban Studies, 51*(4), 655–674.

Martin, G. D. (2018). *The aftermath of redevelopment agencies: a case study on the abolishment of redevelopment agencies and the impact it has had on economic development in the county of Riverside, California*: University of La Verne, (Dissertation/Thesis).

Moore, S. W. (2005). Blight as a means of justifying condemnation for economic redevelopment in Florida. *Stetson Law Review*, 443. Retrieved from http://ezproxy.lib.usf.edu/login?url=http://search.ebscohost.com/login.aspx?direct=true&db=e dshol&AN=edshol.hein.journals.stet35.24&site=eds-live

Morris, E. J. (1974). The Nixon housing program. In (Vol. 9, p. 2): Section of Real Property, Probate and Trust Law / American Bar Association.

Parker, A. C. (2012). Still as moonlight: why tax increment financing stalled in North Carolina. *North Carolina Law Review*, 661. Retrieved from http://ezproxy.lib.usf.edu/login?url=http://search.ebscohost.com/login.aspx?direct=true&db=e dshol&AN=edshol.hein.journals.nclr91.17&site=eds-live

Paull, E. (2008). Using tax increment financing for brownfields redevelopment.

Rast, J. (2012). Why history (still) matters: time and temporality in urban political analysis. *Urban Affairs Review, 48*(1), 3–36.

Reingold, D. A. (2001). Are TIFs being misused to alter patterns of residential segregation? The case of Addison and Chicago, Illinois. *Tax Increment Financing and Economic Development: Uses, structures and impact, State University of New York Press, Albany*, 223–241.

Roberts, P., & Sykes, H. (1999). *Urban regeneration: a handbook*: SAGE Publications.

Schacht, W. (2018). Urban design and city form in redevelopment. In B. Hersh (Ed.), *Urban Redevelopment: A North American Reader* (pp. 42–61). New York: Routledge.

Schilling, J., & Pinzón, J. (2016). The basics of blight. Recent research on its drivers, impacts, and interventions. *VPRN Research & Policy Brief* (2).

Schoettle, F. P. (2003). What public finance do state constitutions allow? *Financing Economic Development in the 21st Century*, 27–49.

Turok, I., & Shutt, J. (1994). The challenge for urban policy. In: Taylor & Francis.

USEPA. (2014). *State brownfields and voluntary response programs*. Retrieved from Washington, D.C.: https://www.epa.gov/sites/production/files/2015- 11/documents/brownfields_state_report_2014_508_12-17-14_final_web.pdf

Weber, R. (2002). Extracting value from the city: neoliberalism and urban redevelopment. *Antipode, 34*(3), 519–540.

Weber, R., & O'Neill-Kohl, S. (2013). The historical roots of tax increment financing, or how real estate consultants kept urban renewal alive. *Economic Development Quarterly, 27*(3), 193–207.

Yadavalli, A., & Landers, J. (2017). Tax increment financing: a propensity score approach. *Economic Development Quarterly, 31*(4), 312–325.

Youngman, J. M. (2011). TIF at a turning point: defining debt down. *Lincoln Institute of Land Policy Working Paper No. WP11JY1*.

8 Balancing the Gentrification Elephant in the Room, or Slaying a Monster? Notes toward a Post-American Urban Future

M. Martin Bosman and Stephen T. Buckman

The term gentrification has become tarred. But called by any other name – revitalisation, reinvestment, renaissance – it would smell sweet.

The Economist (June 21, 2018)

Introduction

There is a currently an urgent academic and policy debate about urban America's deepening social and economic crisis. At the heart of this debate is the problem of gentrification, a process the urban geographer Neil Smith (2002) called the new global urban regeneration strategy. This problem has spurred ongoing political controversy; generated intense academic and popular debate; garnered extensive media and political attention; and reinvigorated urban social movements and civic organizations demanding social and environmental justice (Mendes 2018; Mayer 2010). Indeed, gentrification has become a byword for many of the "most pressing conflict in cities today" (Zukin, in Sevilla-Buitrago 2013: 467).

Gentrification is often a byproduct of the buy low, sell high mindset that drives the real estate industry. As this book intends to show, one of the many justifications of Community Real Estate Development (CRED) is to address the social and economic disparities that gentrification creates by promoting urban environments that are much more inclusive to all members of a community. Yet, good intentions are often the parents of misguided results. Whereas a clear result of CRED is that some communities rebound from an influx of investment capital and economic development, property developers who subscribe to the aforementioned herd mentality of buy low and sell high are usually attracted to previously neglected and underserved areas which results in new 'market rate' (non-CRED) developments which in turn ends in gentrification.

In this chapter, following the call of this special book project, we unpack the following compound question: Is gentrification an anomaly, a regrettable unintended side-effect, of urban redevelopment, or is it in fact its *raison d'être*? Alternatively, is state-enabled and market-driven gentrification,

DOI: 10.1201/9781003109679-9

disguised as 'urban revitalization' and 'neighborhood redevelopment,' an oxymoron? Specifically, is it possible to have revitalization and redevelopment without the negative impacts of gentrification? Or, is gentrification inexorable precisely because it is structurally a "process that urbanizes space at the expense of certain populations" (Zapatka 2017: 228) and because it is historically a "racial project" (Alvaré 2017) that is intended to dispossess and displace various undesirable residents and to appropriate their material and symbolic cultures encoded in the urban built environment as "raw material to accumulate capital" (Burns and Berbary 2020; Misoczky and Misoczky de Oliveira 2018: 1025), using "growth machine[s that are] capable of increasing aggregate rent and ensuring wealth for elites who use the growth consensus to eliminate alternatives" (ibid.)? Simply put, is 'inclusive gentrification' (Ellen 2018) or 'good gentrification' (Cortright 2019) even a possibility given everything we know about the machinations of "actually existing capitalism" (Mann 2013)? In short, is gentrification truly an elephant that defies explanation as the urban geographer Chris Hamnett (1991) once argued, or is it a "monster which is devouring Black communities," as stated by Ron Daniels, president of the national civil-rights network, Institute of the Black World 21st Century, at a three-day event billed The National Emergency Summit on Gentrification in Newark, N.J. in April 2019 (ibw.org)? Indeed, is gentrification a monster: "moving capital from one place to another … in search of profit" as the urban geographer Neil Smith (1992: 113–114) asserted, or is it simply

> a process in which a poor area (as a city) experiences an influx of middle-class or wealthy people who renovate and rebuild homes and businesses and which often result in an increase in property values and the displacement of earlier, usually poorer residents.

as defined by the Merriam-Webster dictionary (merriam-webster.com/dictionary/gentrification)? As these questions imply, the debate about gentrification is as fierce as it is polarizing among urban scholars, city planners, landed interests, and social justice activists.

On the one hand, the proponents of 'good gentrification,' including neoliberal city governments, think-tanks, academics, and elite actors and business professional managing the Finance, Insurance and Real Estate Industrial-Complex (FIREIC, pronounced FREAK), typically insist that gentrification "is usually good news for there is nothing more unhealthy for a city than a monoculture of poverty" (Duany 2001: 1; also see Cortright 2019; *The Economist*, June 21, 2018). They insist that minus a few "unforeseen" issues (Davis 2020), gentrification is basically a benevolent process that is mainly intended to capacitate 'higher and better uses' of land and real estate in "blighted" urban residential neighborhoods and commercial districts in the interest of general prosperity (Herscher 2020). A leading exponent of this view is "the man who reinvented the city" (Redmon, May 18, 2010) and co-founded the New Urbanism movement, Andres Duany (2001: 1), who argued that:

> Gentrification rebalances a concentration of poverty by providing the tax base, rub-off work ethic, and political effectiveness of a middle class, and in the process improves the quality of life for all of a community's residents. It is the rising tide that lifts all boats.

On the other hand, urban social movements, critical scholars, and a new breed of abolitionist planners and activists are busy exposing the "swindle" (Adorno et al. 1950: 678; Marx 1964: 141) of gentrification as a corporate state reaction to resolve the "financial crisis of urbanization" (Harvey, cited in Christophers 2011: 1347) on the backs of the working poor, while justifying their plans as inclusive development and urban revitalization. What this group sees in Andres Duany and New Urbanism's neo-traditional nostalgic vision of cities is the making of a new accumulation strategy of "gentrification by ground rent dispossession" (Lopez-Morales 2010), especially in impoverished and disenfranchised communities of color and class. For instance, Neil Smith, a leading critic of gentrification, saw gentrification as a vicious project of class revanchism. "The rallying cry of the revanchist city might as well be: 'Who lost the city? And on whom is revenge to be exacted,'" was how he put it (Smith 1996: 227). In this light, despite its putative intentions and rosy rhetoric, gentrification ought to be considered a strategy by corporate and state elites to "dispossess an increasingly interconnected and global urban citizenry" (Zapatka 2017: 229) as a way to resolve the "capitalist crisis of over-accumulation through its role in absorbing surplus capital" (Christophers 2011: 1347). The debate about this extant global urban project of accumulation by dispossession and displacement as a response to the structural crises of capitalism has archetypically proceeded along class (e.g., see Smith 2007, 2002, 1996; Slater 2014, 2006) and race lines (e.g., see Lipman 2018; Roy 2018; Stovall 2021, 2016). However, before we reviewed this ongoing debate, a brief definition of gentrification might be useful.

CRED as Gentrifier

As the other chapters in this book point out, CRED's focus is on promoting the communitarian and civic dimensions of real estate development and not just pursuing orthodox market-driven goals. In this regard, the policy objective is to entice developers to engage in community-oriented development projects by enabling them to balance out their losses as a result of having to reserve at least 20% of their projects for affordable housing and other community-oriented services. To make this a reality, various government incentives have been introduced. For instance, Chapter 4 outlines the major tax incentives that have been introduced in recent years to incentivize

CRED, including Low Income Housing Tax Credits (LIHTC), New Market Tax Credits (NMTC), and Opportunity Zones. While the verdict is still out regarding the relationship between NMTCs and Opportunity Zones, on the one hand, and gentrification, on the other, research shows that LIHTC have a strong relationship with gentrification.

The main problem with these tax incentives, especially LIHTC, is the "contradiction between rhetoric and reality" (Layser 2019: 1). Specifically, they "are presented as laws that benefit low-income communities, yet the dominant types of place-based investment tax incentives are not designed for this purpose" (ibid.). Moreover, tax credit incentives are administered federally and incentivize individual buildings rather than entire neighborhoods, which benefit private developers more than communities (McCabe 2019). A property tax credit may also stimulate housing supply and cause changes in amenities, thereby attracting more for-profit businesses, which eventually affects rental prices (Baum-Snow and Marion 2009). A study by Benjamin Field (2016) showed that investors are more likely to invest in gentrifying neighborhoods as the probability of getting tax credits is higher.

While LIHTC and investment tax incentives of this nature may help some low-income individuals get housing, they also displace poor people who cannot afford the new housing stock. For instance, Brian J. McCabe (2019) found that these investment tax incentives return abandoned or neglected buildings to the for-profit market which leads to higher housing prices and limited housing opportunities for low-income households, depending on the characteristics of each neighborhood. This finding is consistent with a study by Joanna R. Shell (2020) which showed that the number of low-income housing tax credit properties is associated with gentrification in the neighborhoods. While Shell's study emphasized the benefits of gentrification, including increased amenities and other physical resources in neighborhoods, the study also showed that these incentives often lead to displacement and changes in the socioeconomic and demographic character of the neighborhoods. Among the more noticeable changes are increasing in the number of white homeowners and renters and higher housing and rental prices (Shell 2020; also see Field 2020; McCabe 2019). However, these findings are at odds with studies by Lance Freeman (2016, 2005, 2004, 2002) which insist that gentrification causes displacement in some rare cases.

If the supposedly good intentions of community-oriented policies such as CRED still result in gentrification, the question then becomes: what exactly constitutes gentrification and how did we get here? The remainder of this chapter will look at what gentrification is from three areas of inquiry. The Marxist, race, and neoliberal traditions of thought with particular emphasis on the neoliberal tradition which is most closely aligned with the real estate industry (see Bosman and Buckman (forthcoming) for a more detailed analysis of all three).

What Is Gentrification?

> [T]his social and economic process of neighborhood restructuring has become one of the most controversial challenges facing cities in the twentieth century.
>
> Stacey Sutton (2020: 66)

Ever since the U.K. sociologist Ruth Glass (1964) coined the term, gentrification has metastasized into a planetary process (Lees et al. 2016). In fact, it has spread with such speed and ferocity that we can now confidently declare that 'global society is well on its way to being completely gentrified' (see Lees et al. 2016; Wyly 2015; Smith 2002). In fact, some have gone so far as to claim that all contemporary "urbanization [is] gentrification" (Zapatka 2017: 228). Most Anglo-American scholars still trace the term's origins to postwar London where Glass first noticed the displacement of working-class communities from inner-city boroughs like Islington and Notting Hill and the transformation of their "shabby, modest mews and cottages" into "elegant, expensive residences" (ibid: xvii). This is how Glass described this transformation:

> One by one, many of the working-class quarters have been invaded by the middle class – upper and lower ... Once the process of gentrification starts in a district it goes on rapidly until all or most of the working-class occupiers are displaced and the whole social character of the district is changed.
>
> (ibid.)

"London," she added, "may quite soon be a city which illustrates the principle of the survival of the fittest – the financially fittest, who can still afford to work and live there;" moreover, "London may soon be faced with an *embarras de richesse* in her central areas" (ibid: xx). And, as "London becomes 'greater,' the dislike, indeed, the fear, of the giant grows" (ibid: xxvi). Glass called this early process of real estate capital invasion and destruction of working-class neighborhoods by the English *gentry* who existed beneath the nobles in Britain's medieval caste system. Like the nobles, the *gentry* lived off the rents extracted from tenant farmers working their ancestral lands while they governed the countryside as magistrates and legal trustees of their tenants. Given this changing economic landscape, large landowners with more productive tenants eventually prevailed, displacing their relatively poorer and less competitive peers, many of whom repaired to London's burgeoning industrial and real estate markets in the 19th century to engage in "petit-bourgeois entrepreneurial and rentier activity" (Bridge 1995: 236).

Many aspects of this formative process of urbanization, or "capital switching" (Harvey 2009, 2008, 1985), which resemble a form of 'primitive,' incipient gentrification accelerated and mutated throughout the long postwar economic booms and busts that defined the early 20th century. By the time Glass studied the transformation of inner-city London, many of these processes were already well entrenched. As we will show later on, there are alternative historiographies of the causes of gentrification. For now, suffice it to say that "gentrification is one of the most significant and socially unjust processes affecting cities worldwide today," according to a group of prominent urban theorists (Lees et al. 2016, no page). But the process has materialized along different conjunctural planes in different times and places (Lees et al. 2016; Bridge 1995: 237). Accordingly, it signifies different things to different people in different cities within the power geometries of global capitalism. For instance, recent studies show that in different cities, gentrification is either associated with "slum clearance" (Ascensão 2015; Ghertner 2014), or "mega-events" like the Olympics and the World Cup (Clark et al. 2016; Müller 2015), especially in societies governed either by strong neoliberal or developmental states with weak and/or authoritarian regimes. On the other hand, at the city scale, for example, in places like Portland, Oregon, where gentrification is de facto public policy, it is justified as a natural and inevitable outcome of consumer choices and free-market forces. As one Portland developer explained: "We always think it's a somebody, and in my opinion it's an economic force – there's no one orchestrating this outcome" (Sanneh, July 4, 2016).

Gentrification's variable instantiations have fueled spirited debates and competing theories about its causes and consequences for cities and residents (e.g., see McElroy and Werth 2019; Smart and Smart 2017; Ghertner 2015, 2014; Moulatas 2011). One result of this debate is that gentrification has morphed into something of an academic laboratory that seems to continually discover new virulent strains, including "third-wave gentrification" (Hackworth and Smith 2001); "super-gentrification" (Lees 2003); "mega-gentrification" (Lees 2011); "planetary gentrification" (Lees et al. 2016); and "hyper-gentrification" (Moss 2017), to name a few recent variants. Given the limitations of space, however, we will conduct a full review of the range of gentrification variants and their significance for current policy and practice (see Lees et al. 2016, 2008; Atkinson and Wulff 2009 for comprehensive reviews). Instead, we offer a partial overview of what we consider the three most dominant and conflicting approaches to gentrification in the U.S. research literature.

Marxist Framing of Gentrification as a Class Project

I think it is obvious that in a capitalist society one's preferences are more likely to be actualized, and one can afford grander preferences, to the extent that one commands capital. We may regret that economics so strongly affects one's ability to exercise preferences, but it would hardly be prudent to deny it; preference is an inherently class question. The preference to invest capital in, for example, inner city rehabilitation and redevelopment, in search of profit, is a powerful social force.

Neil Smith (1992)

As the quote above from a prominent urban geographer, Neil Smith, suggests, this class school of thought, consisting mostly of Marxisant urban geographers and other critical urban scholars, maintains that despite the lofty win-win rhetoric of its boosters, gentrification, as the "vanguard of the new post-industrial, or even post-modern city" (Byrne 2003: 407), is structurally and institutionally a class revanchist project that caters to the long-term strategic financial interests of landowners, corporate rentiers, and fiscally-strapped municipalities (Slater 2006; Smith and deFilippis 1999; Smith 1979, 2002), and secondarily to the aesthetic and consumption habits and interests of "middle class professionals and managerial workers, cultural intermediaries and new 'expressive' professionals" (Wynne 1998: 183; also Baum-Snow and Hartley 2016; Couture and Handbury 2015; Zukin 2010, 1995, 1989; Ley 1980, 1996). Furthermore, class theorists argue that to fully explain the structural drivers of gentrification, scholars, policymakers, and communities must be attentive to shifts in the processes and institutions that shape the uneven geographies of capitalism (Brenner 2014; Slater 2006). This requires an understanding of Marxist political economy to elucidate how the creative-destructive proclivities of capitalist uneven development foreshadowed the various trajectories of neoliberalization and gentrification since the 1970s (Lefebvre 1991, 2003; Harvey 1985; Smith 1984).

Thus, the transformation of U.S. cities into postindustrial, financially-led growth machines, gentrification emerged as a "major vehicle of capital accumulation" (Slater 2012: 189). This involved sacrificing poor racialized neighborhoods by designating them as 'blighted,' justifying the use of eminent domain to either expropriate or existing property and treating poor legacy residents as disposable workers (Povinelli 2011; Wilson 1987) and as expendable, "defective and disqualified consumers" (Bauman 2009: 2). Over time, more and more racialized neighborhoods were officially discounted as places with little to no intrinsic value and, therefore, "rendered lootable" (Wang, cited in Bu 2018) for the purposes of revalorization by corporate investors, property developers, financial speculators, PMCs, rich empty nesters, conspicuous consumers, and well-off tourists. As such, gentrification "expresses the impulses of capitalist production" and conspicuous consumption in the service of economic and political elites (Smith 2002: 427) "rather than social reproduction" (ibid.) of poor racialized communities and households (Howland 2020). In the process, without a trace of racial irony, mostly white bankers and private financial speculators with over-accumulated money capital presented themselves as 'urban pioneers' who were dedicated to the revitalization of mostly poor black and brown inner-city neighborhoods into safe, desirable, and profitable 'work, play and stay destinations' for a new urban gentry of PMCs and propertied classes, as well as privileged consumers, and wealthy tourists (see Katz and Wagner 2014).

Given the globalization of the capitalist crisis and urban restructuring, gentrification is also global in scope, enclosing New York City and San

Francisco (Bianco et al. 2018; Casque 2013); London and Berlin (Holm 2013; DeVerteuil 2011); Johannesburg and Cairo (Wahba 2020; Goo 2018); Mexico City and Buenos Aires (Janoschka and Sequera 2016); and Manilla and Beijing (Ortega 2016; Shin et al. 2015), to name a few. Instead of being seen as organized efforts of urban capital accumulation and rent extraction, these cities have been promoted as icons of global urban development and inclusive growth by the corporate media, neoliberal think-tanks, and corporate universities (see Florida 2017, 2005, 2002) Meanwhile, studies show that gentrification has enabled "a new medievalism that separates and segregates rich and poor communities" (Gill 2003: 201), and that these "bright archipelagos of utopian luxury and 'supreme lifestyle' are mere parasites on a planet of slums" (Davis 2007: 4). In other words, from a class perspective, gentrification has produced a new global geography of capital accumulation and economic inequality that is steadily enclosing poor urban neighborhoods like Wynwood, Little Haiti, and Overtown in Miami-Dade, the Mission District and South of Market in San Francisco, and Brooklyn and Harlem in New York City.

These results of this process are so consistent that it is reasonable to assume that displacement, dispossession, and eviction are intentional features of gentrification. However, the corporate and rentier elites who promote the providence of gentrification continue to conflate its pro-growth promises and anti-poverty rhetoric about building better and more inclusive cities with its dismal failures to realize the goal of urban "development as freedom" (Sen 1999). This is how one critic explained this neoliberal conflation:

> Here is how gentrification talk typically goes: poor neighborhoods are said to need 'regeneration' or 'revitalization,' as if lifelessness and torpor – as opposed to impoverishment and disempowerment – were the problem. Exclusion is rebranded as creative 'renewal.' The liberal mission to 'increase diversity' is perversely used as an excuse to turn residents out of their homes in places like Harlem or Brixton – areas famous for their long histories of independent political and cultural scenes. After gentrification takes hold, neighborhoods are commended for having 'bounced back' from poverty, ignoring the fact that poverty has usually only been bounced elsewhere.
>
> (Madden, October 10, 2013)

Black Radical Tradition: Gentrification as Both a Race and Class Project

> [A]s the history of urban renewal, gentrification, and urban expressway building shows, central city disinvestment and reinvestment are part of the dynamics of urban race relations, with black and other minorities having to bear the brunt of displacement and neighborhood displacement.
>
> Kevin Fox Gotham (2001)

Not surprisingly, gentrification has become a rallying cry among poor ra-cialized communities across urban America (Institute of the Black World 21st Century, April 2019; Romano and Franke-Ruta 2018). A simple reason for this development is that gentrification has thrown into sharp relief the slow sociospatial violence that asymmetrically targets poor racialized com-munities that reside in or close to financially and culturally desirable but his-torically underinvested and underserved neighborhoods (Naram 2017). This contradiction brings us to an allied school of gentrification studies that is inspired by the "radical black tradition" (Robinson 2000 [1983]). This school considers class theories useful, but too colorblind because they fail to ac-count adequately for the historically and institutionally entrenched racial and gendered logics of gentrification (Danewid 2020; Ramirez 2020; McEl-roy and Werth 2019; Mumm 2017; Lees 2000). It holds that absent major qualifications, class theories are unable to fully explain the intimate links between gentrification and the disproportionate dispossession and displace-ment of poor black and brown communities (e.g., see Lewis 2020). Thus, this school argues that class theories tend to underestimate the changing but enduring role of "race (as deployed by racism)" (McMillan Cottom 2020: 443) and the "racialization of space" (Zimmer 2020; Lipsitz 2011; Mills 1997: 2–3) that are co-constitutive of gentrification (Mumm 2017, 2014). In short, this school is based on the premise that gentrification is not just a function of "accumulation by dispossession" (Harvey 2003) of an abstract, deracinated working-class communities, as implied by class theories. Rather, it is funda-mentally a function of "racialized accumulation by dispossession" (Wang 2018: 99–150) that is predicated on historical constellations of public and pri-vate institutions and practices of racialization and negative discrimination which point to "how the logics of differentiation mediate capitalist accumu-lation" (Wang, cited in Buna, May 13, 2018).

Dominant Neoliberal Framing of Gentrification: An Unintended Byproduct of Community Redevelopment

> All we want to do is bring value to an abused area. At the end of the day, we don't feel like we're coming in and actually trying to displace anybody. I get it comes across this way … It does change the price point from what it was to it becomes, but that's just the natural growth of the city. It's going to happen at some point. All we can do is make it as easy a transition as possible.
>
> Agent for Luxury Condo Development,
> Seattle (Glenn Nelson, November 2019)

This brings us to the final school of thought and the one most clearly con-nected and of concern to CRED, namely, the real estate-cum-urban revi-talization school. This school rejects both the radical black tradition and orthodox class theories of gentrification. The adherents of this compara-tively tiny yet powerful "neoliberal thought collective" (Mirowski and

Plehwe 2009) include international investors, government technocrats, bankers, developers, realtors, lawyers, planners, architects, land- and property owners, to name a few beneficiaries and sinecures. Against class theorists who insist that "gentrification is a back-to-the-city movement of capital, not people" (Smith 1996, 1979), adherents of this school retort that "gentrification [is] simply the result of peoples' choices expressed through the market" (Godsil 2014) and/or "shifting demographics, such as the aging population (i.e., empty nesters without children) and changing preferences for high-amenity locations like downtowns" (Hartley et al. 2016: 139; also see Buzar et al. 2007a, 2007b). Contesting a vast body of empirical evidence, and donning "masks of positivity" (Burns and Berbary 2020: 2), they cling to presumptions of 'public good' that undergird gentrification discourses by insisting that the:

> gentrification process reverses decades of urban decline and could bring broad new benefits to cities through a growing tax base, increase socioeconomic integration, and improved amenities … [and] … has important benefits for low-income residents, such as improving the mental and physical health of adults and increasing the long-term educational attainment and earning of children … Gentrification thus has the potential to dramatically reshape the geography of opportunity in American cities.
>
> (Brummet and Reed 2019: 1)

Other adherents of this school have denied outright the very existence of gentrification. This attitude of this group recalls a witticism by the French poet Charles Baudelaire (1864) that: "the cleverest ruse of the devil is to persuade you he does not exist" (Smith 1919: 82). Today, the devil's or monster's ruse that is popular among neoliberals is to deny that gentrification exists by dismissing it as an "urban myth" (e.g., see Freeman, June 3, 2016; Matthews, July 29, 2016; Buntin, January 14, 2015) despite compelling evidence that is a central accumulation strategy among financial and rentier capitalists (Hudson 2020; Shaw 2020; Slater 2017). A variation of this ruse is to minimize the existence and impact of gentrification on the grounds that "it's risky: too many implications, too many donors who might not like what they hear," according to Daniel Squadron, a former New York State (cited in Gibson 2015; also see Glazer, September 6, 2019; Barr 2018).

Behind this devil's ruse lurks a matrix of predatory institutions, laws, policies, programs, and practices that either ignore, deny, or minimize the "roots of structural deprivation and places culpability on Black individuals' assumed cultural deficiencies for the state of dilapidated inner-city neighborhoods" (Alvaré 2017: 116). And although it should be obvious that gentrification is structurally incapable of delivering the new 'geography of opportunity in American cities' its proponents profess, this school's most ardent supporters and practitioners continue to command enormous influence within the ideological apparatuses of the state (Althusser 1971), including in educational institutions like universities, think-tanks, and urban consultancies (Baldwin 2021); the legal and planning professions (Godsil May 2014); corporate media and digital platforms (Cortright, March 10, 2020; Cortright, August 28, 2018; *The Economist*, June 21, 2018; Freeman, June 3, 2016); corporate boardrooms and financial agencies; and national and global institutions responsible for global finance and investment like the U.S. Federal Reserve (Su 2019; Ding et al. 216) and World Bank (Amirtahmasebi et al. 2016). With typical tendentiousness, these interest groups continue to peddle the "neoliberal swindle" (Freedman 2019) that gentrification is an unimpeachable public good whose negative racial impacts are unfortunate but ultimately unintended and inconsequential (e.g., see Capps, July 16, 2019; *The Economist*, June 21, 2018, February 21, 2015; Millsap, March 29, 2018; Wynn and Deener, October 10, 2017; Matthews, July 29, 2016; Byrne 2003; Duany 2001). As stated by the Seattle-based agent for a luxury condominium project in the opening epigraph to this section, gentrification is presented to skeptical communities as a process that operates more or less according to the natural principles and timeless objectives of individual choice and free markets. As such, it is presented as a facially race- and class-neutral response to urban poverty that unintentionally disadvantages some, but ultimately benefits everyone because it is 'making America's cities great again' as they "vie for investment, talent and business" in the context of global inter-urban competition, according to the CEO of Citi Bank, Vikram Pandit (European American Chamber of Commerce, March 13, 2011). In a nutshell, the gentrification swindle is: 'don't worry that we're destroying your neighborhood and scattering your community to the winds. It's going to be good for you in the long run.'

As mentioned earlier, the FIREIC's belief that the salvation of urban America lies in the providence of gentrification seems almost religious despite mounting evidence that the "real estate state" (Stein 2019) is primarily focused on serving the interests of financial and rentier capitalists, and secondarily those of the PMC and effluent residents, consumers, and tourists, while justifying the gentrification of inner-city neighborhoods in the name of public good. The ideological resilience and rhetorical dexterity of this view are so hegemonic in a Gramscian sense that it is "now the common-sense way we interpret, live in, and understand the [urban] world" (Harvey 2007: 22). It is a resilience and dexterity that is invariably dependent on and mediated through the power of two of the most enchanting ideas (Carlsten 2019: 3) in neoliberal urbanism's playbook (e.g., see Slobodian 2018; Eagleton-Pierce 2016), namely, 'redevelopment' and 'regeneration' or 'revitalization' (Burns and Berbary 2020; Lovering 2007). Here, it is worth noting that the provenance of these racially coded ideas stretches back to an earlier Anglo-American playbook and the post-World War II neocolonial argot about the 'development' of 'backwards' Third World nations and regions (Pieterse 2010, 2000; Escobar 2000, 1995; Cooper 1997; Crush 1995). Contemporary permutations of modernization theory and practice are infested with the same historical conceits about development and colonizing tropes about social cleansing and spatial purification (Wacquant 2008: 198; Wilson 2004: 774; Swanson 2007; Smith 2001, 1996: 26–27) of "blighted' urban neighborhoods" (Herscher 2020) through the introduction of "green urban amenities" (Anguelovski et al. 2019; Bunce 2019; Black and Richards 2002); the innovation of "smart cities" or "universities" (Baldwin, July 30, 2017) with "resilient infrastructures" (Spinney 2021; Beretta 2018; van den Bosch 2017); the building of climate-friendly utilities and carbon-neutral services (Rice et al. 2019); and the opening of exciting new entertainment venues in underinvested and underserved inner-cities (van der Hoeven and Hitters 2020; Hollands and Chatterton 2003).

Many contemporary gentrification schemes are riddled with racist assumptions about how the introduction of a new urban civilizing mission via the influx of new investors, residents, and consumers, bringing new work habits, urban goods, services, amenities, etc. will make poor neighborhoods conducive to capital investment which in turn will trickle down to legacy residents and thereby trigger "the rising tide that lifts all boats" (Duany 2001). Thus, based on this virtuous cycle: "on balance, by increasing the number of residents who can pay taxes, purchase local goods and services, and support the city in state and federal political processes" (Byrne 2003: 406), everyone will benefit, including "the poor and ethnic minorities" (ibid.). A key presumption is that by dispersing and/or integrating spatially concentrated pockets of black and brown poverty and attenuating their cultural stigmatization and political disenfranchisement, gentrification will improve their lot (Freeman and Braconi 2004, 2002) by increasing their well-being, security, and happiness (see Cortright 2018; Brodeur and Flèche 2017) and thereby "enhance the political and economic positions of all" (Byrne, ibid.). What this "Holy Grail of urban competitiveness" (Lovering 2007: 344) that is considered a natural artifact of gentrification refuses to accept is that unless there is a fundamental break with "racial capitalism" (Kelley 2017; Melamed 2015), the uneven and inequitable production, distribution, and consumption of urban space will continue to exacerbate preexisting social injustices because state-subsidized and market-led urban revitalization "works in peculiarly seductive ways with whiteness because of the way it seems rooted in commonsense" (Haymes 1995: 104).

This racist ontology of land and property values subtending capitalism that equates "white places with privilege, from the neighborhood and race effects that create unequal and unjust geographies of opportunity" (Lipsitz 2011: 28), combined with the neoliberal imaginary of "hostile privatism, defensive localism, and competitive consumer citizenship" (ibid: 124), continues to inspire gentrification's public and private investors, enforcers, and beneficiaries, including city governments; bankers and real estate investment trusts (REITs); developers, realtors, urban planners, and universities; and other elite sinecures who profit off the $217 trillion FREIC that accounts for 60% of global assets (Schloredt, February 24, 2020). Collectively, this complex operates what the urban geographer Ash Amin (2013: 476) called "'business consultancy' urbanism" (e.g., see Brummet and Reed 2019; Ellen 2018; Florida and McLean 2017; Edlund et al. 2016; Nevius 2013; Vigdor 2010; Freeman and Braconi 2006; Byrne 2003; Vigdor et al. 2002).

This entrenched racialized neoliberal imaginary and its "[t]alk of 'urban regeneration' is intended to convey an impression that something new is happening, which is yet at the same time a return to something old and valuable" (Lovering 2007: 344). This agenda is promoted by other FIREIC stakeholders, including the corporate media and middle- and upper-class social-media influencers (see Cortright, March 10, 2020; Kasdin 2020; Hakimi, March 2019; Cortright, August 28, 2018; Freeman, June 3, 2016; Ray, May 25, 2016; Buntin 2015; Hymowitz, November 13, 2015; Star Tribune Editorial Board, August 19, 2015; Davidson 2014; Sullivan 2014; Hopkinson, September 9, 2012). This fraction of FIREIC has continued to push the aesthetic appeal of gentrification by turning it into powerful counternarratives that render poor people who have problems into people who are problems, to paraphrase George Lipsitz (2018).

To that end, a popular strategy to normalize gentrification as unproblematic and race-neutral is to frame it through the corporate media's time-honored 'two sides to every story' and 'fair and balanced' lens. This is an extremely popular strategy of symbolic violence that tries to naturalize and depoliticize the many injustices accompanying gentrification. Briefly, the strategy provides equal media time and space to pro- and anti-gentrification arguments under the pretext of reportorial 'balance' between 'partisan' interests vying for public legitimacy (e.g., see O'Neil, February 11, 2019; Martin, March 12, 2014; Mangona, July 10, 2012). Based on this equivocal framing and bogus symmetry, gentrification is presented to the public as both good and bad, implying that it ultimately comes down to personal choice among gentrifiers and gentrified under voluntary and free-market conditions between a moribund status quo, meaning, increasing neighborhood disinvestment and urban decay, and a new dawn, meaning, gentrification and urban renaissance (e.g., see Cortright, March 10, 2020; *The Economist*, June 21, 2018; Hopkinson, September 9, 2012, among others). Either choice has its costs and its benefits, the public is told. Based on this false equivalence and both-side-ism which permeates the corporate media and neoliberal scholarship, mountains of incontrovertible *evidence* that gentrification leads to racialized dispossession and displacement is either rendered superficial or accidental, or presented to the public as debatable by oceans of corporate media *propaganda* that strenuously denies that gentrification is a necessary precondition for recapitalizing cities in 'free-market' democracies like the United States and thus not worthy of government intervention and public censure. The urban geographer Tom Slater (2014) called this deliberate two-sided framing of gentrification "false equivalence urbanism," because it takes empirical evidence of displacement and reduces it to merely one of many plausible narratives. The FIREIC and local authorities continue to use this reportorial sleight of hand to legitimize the necropolitics of gentrification and to delegitimize the politics of imagining just alternative urban futures as demanded for by urban abolitionist groups like Critical Resistance, an Oakland, California-based grassroots organization that contests dominant representations of public safety; Black Visions, a Minneapolis-based community network that organizes around city budgets; BYP100 (Black Youth Project 100), a national member-based organization that mobilizes around local issues; INCITE!, a network of radical feminists of color that organizes to end state, corporate, and patriarchal violence in homes and neighborhoods; and of course Black Lives Matter (Issar 2020), among other urban reform and anti-systemic groups and movements.

What Are the Solutions?

> I could not have anticipated among all this urban growth and revival that there was a dark side to the urban creative revolution [read: gentrification], a very deep dark side. The urban pessimists have a point. We neglected their point, which is that cities are gentrifying, people are being priced out, displaced from their homes.
>
> Richard Florida (2017)

Since the benefits of gentrification are not as self-evident as its supporters allege (Brummet and Reed 2019; Cortright 2019; *The Economist*, June 21, 2018) and the costs are so increasingly evident that it has been vilified as a project of "new urban colonialism" by some (Danewid 2020; Mitchell 2019; Quizar 2019, 2020; Jackson 2017; Urena-Ravelo 2017; Right to the City Montreal 2012; Wharton 2008; Atkinson and Bridge 2005; Smith 2005: 15) and "new Jim Crow" by others (J.T. The LA Storyteller 2020; Wagner 2017), the majority of our efforts will have focus on finding long-term structural solutions instead of peddling short-term fixes about "transforming gentrification into integration," as advised by the Seton Hall University law professor Rachel D. Godsil (May 2014). In an age that is dominated by the "madness of economic reason" (Harvey 2017), the time for balancing the elephant of gentrification is surely over. Instead of returning to the "camp thinking" (Hill Collins 2002; Gilroy 2000; Morrison 1993) that has shaped gentrification for more than a generation, why not just slay the monster instead of using its mounting crises as neoliberal alibis to continuously reformulate new promises of enhancing the rights to the city of everyone regardless of race, class, gender, or any other social hierarchies? If the "undesirable social impacts" of gentrification (World Bank, no date) can no longer be 'balanced' by including sections of the racialized poor in the spoils of gentrification (Cortright 2019) as implied by some proponents of more inclusive gentrification (Ellen 2018), then a radical rethinking of urbanization and city-building would seem to be the only just and democratic way forward.

This then leads to this inevitable question: "What is to be done? And who the hell is going to do it?" (Harvey and Wachsmuth 2012: 264)? Fortunately, a growing number of critical scholars and community leaders are already imagining and, in some cases, building new urban frameworks beyond the dominant paradigm reviewed above. However, if the two major sectors and institutions at the heart of America's contemporary urban crisis, namely, the state, and finance, insurance, and real estate, are serious about fostering authentic 'community development as freedom,' as defined by Amartya Sen (1999), one of the first steps would be to center the basic rights and needs of poor and working communities in urban redevelopment. Moreover, the state and the FIREIC will have to be compelled by progressive urban social movements to accept an equitable redistribution of economic resources and to respect the full and meaningful empowerment of the urban poor and working masses. It is only once the entrenched historical systems of expropriation and relations of economic exploitation, the institutions of political domination, and the cultural hegemony of white supremacy that equate wealth with whiteness and "whiteness" with the "ownership of the earth forever and ever" (Du Bois 1920: 29) have been abolished that those interested in genuine urban revitalization can entertain discussions about the advisability of this technocratic policy or that bureaucratic program. For that to occur, however, a significant if not the: "entire segment of white society would have to step outside the white class structure and become, in some unforeseeable sense, ex-white" (Martinot 2003: 204). And for that to happen, America's working and proletarian masses would have to demand "an end to the property entitlement of whiteness, and thus the property basis of society itself" (ibid.: 208). This in turn would require that "the white corporate state that concretizes [whiteness] would have to be demolished" (ibid.). Anything less will continue to prolong the cycles of dispossession, displacement, and eviction of poor communities of class and color from their neighborhoods and other economic and political space of power and representation.

Resolving the growing contradictions and crises of urbanization that gentrification has produced is neither a case of simply including more socioeconomic groups and neighborhoods in the economic spoils as some have implied (e.g., see Brown-Saracino 2009; Patillo 2007), nor is it a case of "ignor[ing] concentrated poverty by pushing it elsewhere" (Powell and Spencer 2003: 442). If gentrification is ultimately about economic resources, ideological hegemony, and political domination, and their continuing institutional entanglements with the privileges and entitlements of whiteness and white supremacy (Lipsitz 2006 [1998]), then nothing short of an urban revolution is required. And if, as Jeremiah Moss (2018: np) has argued: "gentrification is about class – and the places where class interests intersect with race and other factors … but it always about an imbalance of power. And in every scenario, the gentrifiers have more power," then again nothing but an urban revolution will do. Thus, to overcome generations of inequalities in economic and political power and their imbrication with racism and sexism, among other social hierarchies, community activists, insurgent urban planners, and their allies within academia and other state and civic institutions must draw on the theories, policies, and practices of global, postcolonial urban movements that center and prioritize the decommodification and deracialization (and eventual decolonization) of land, real estate, and other related resources so as to address and redress the historical inequalities between rich and poor communities and neighborhoods instead of contributing to the production and normalization of new urban spaces for "[t]ransnational and local real estate investors and financial speculators to zero ... in on" (Lipman 2018: 3) and thus sacrifice "decapitalized inner-city areas for a new round of investment – pushing out the people who lived there" (ibid.). As the urban geographers Kathe Newman and Elvin K. Wyly (2006: 31) put it, as a matter of public policy, the demands of the racialized poor for the right to the city will and can only be won when "the use value of neighborhood and home" is allowed to trump the "exchange values of real estate as a vehicle for capital accumulation." Given the centrality and tenacity of American racism, however, this will only happen when the historically dominant habits of coupling and associating land and real estate values with 'race' (Alcorn, April 20, 2021; Gudell, June 15, 2020; Perry, December 7, 2018; Mumm 2017) are outlawed and criminalized. As a matter of practical politics, this means that insurgent planners and social activists must rally in support of grassroots organizations and those impoverished communities that are most vulnerable to the colonizing powers of gentrification. It is only when the rights of poor and working-class communities to the contemporary city are guaranteed by the injunctive power of the law that the rights of everyone else will also be guaranteed.

Works Cited

Adorno, Theodore, W., Else Frenkel-Brunswik, Daniel J. Levinson, and R. Nevitt Sanford, eds. 1950. *The Authority Personality.* New York: Harper and Row.

Alcorn, Chauncey, April 20, 2021. Homes in Black Neighborhoods Are Undervalued by $46,000. *CNN.* https://www.cnn.com/2021/04/20/economy/redfin-housing-boom-race-discrimination/index.html

Alvaré, Archer Melissa, 2017. Gentrification and Resistance: Racial Projects in the Neoliberal Order. *Social Justice* 44(2–3): 113–136.

Amin, Ash, 2013. Telescopic Urbanism and the Poor. *City* 17(4): 476–492.

Amirtahmasebi, Rana, Mariana Orloff, Sameh Wahba, and Andrew Altman, 2016. *Regenerating Urban Land: A Practitioner's Guide to Leveraging Private Investment.* Washington, DC: World Bank.

Anguelovski, Isabelle, James, J.T. Connolly, Melissa Garcia-Lamarca, Helen Cole, and Hamil Pearsall, 2019. New Scholarly Pathways on Green Gentrification: What Does the Urban 'Green Turn' Mean and Where Is It Going? *Progress in Human Geography* 43(6): 1064–1086.

Ascensão, Eduardo, 2015. Slum Gentrification in Lisbon, Portugal: Displacement and the Imagined Futures of an Informal Settlement. In Lees, Loretta, Hyun Bang Shin, and Ernesto López-Morales, eds. *Global Gentrifications: Uneven Development and Displacement.* Bristol: Policy Press: 37–58.

Atkinson, Rowland and Gavin Bridges, eds. 2005. *Gentrification in a Global Context: The New Urban Colonialism.* London: Routledge.

Atkinson, Rowland and Maryann Wulff, 2009. *Gentrification and Displacement: A Review of Approaches and Findings in the Literature.* Positioning Paper No. 115. Southern and Monash Research Centers: Australian Housing and Urban Research Institute.

Baldwin, Davarian, L., 2021. *In the Shadow of the Ivory Tower: How Universities Are Plundering Our Cities.* Hatchett: Bold Type Books.

Baldwin, Davarian, L., July 30, 2017. When Universities Swallow Cities. *Chronicle of Higher Education.*

Baldwin, Davarian, L., April 29, 2021. BAR Book Forum: Davarian Baldwin's 'In the Shadow of the Ivory Tower.' Edited by Roberto Sirvent. *Black Agenda Report.* https://www.blackagendareport.com/bar-book-forum-davarian-baldwins-shadow-ivory-tower

Barr, Jason, M., 2018. *Bad? Building the Skyline: The Birth & Growth of Manhattan's Skyscrapers.* New York: Oxford University Press.

Bauman, Zygmunt, 2009. *Does Ethics Have a Chance in a World of Consumers?* London: Harvard University Press.

Baum-Snow, Nathaniel and Daniel Hartley, 2016. Causes and Consequences of Central Neighborhood Change, 1970–2010. Paper presented at the Research Symposium on Gentrification and Neighborhood Change, May 25, 2016. Philadelphia: Federal Reserve Bank of Philadelphia.

Baum-Snow, Nathaniel, and J. Marion, 2009. The Effects of Low-income Housing Tax Credit Developments on Neighborhoods. *Journal of Public Economics* 93 (5–6): 654–666.

Beretta, Ilaria, 2018. The Social Effects of Eco-Innovations in Italian Cities. *Cities* 72: 115–121.

Bianco, Federica, Karen Chapple, Neil Kleiman, Stanislav Sobolevsky, Dana Chermesh Reshef, Hao Xi, Gerardo Rodriquez Vasquez, and Ruben Hambardsumyan, 2018. *Map of Gentrification and Displacement for the Greater New York.* NYU Center for Urban Science and Progress. http://www.udpny.org/static/media/report.8f3f1564.pdf

Brenner, Neil, ed., 2014. *Implosions/Explosions: Towards a Study of Planetary Urbanization.* Berlin: Jovis Verlag.

Brodeur, Abel and Sarah Fleche, 2017. Neighbors' Income, Public Goods and Well Being. Working Paper, University of Ottawa.

Brown-Saracino, Japonica, 2009. *A Neighborhood That Never Changes: Gentrification, Social Preservation, and the Search for Authenticity.* Chicago: University of Chicago Press.

Black, Katie, Jo and Mallory Richards, 2020. Eco-Gentrification and Who Benefits from Urban Green Amenities: NYC's High Line. *Landscape and Urban Planning* 204: 1–14.

Bridge, Gary, 1995. The Space for Class? On Class Analysis in the Study of Gentrification. *Transactions of the Institute of British Geographers* 20(2): 236–247.

Brummet, Quentin and Davin Reed, 2019. *The Effects of Gentrification on the Well-Being and Opportunity of Original Resident Adults and Children.* Working Papers 19–30: Federal Reserve Bank of Philadelphia.

Buna, M. May 13, 2018. Carceral Capitalism: A Conversation with Jackie Wang. *LA Review of Books.* https://lareviewofbooks.org/article/carceral-capitalism-conversation-jackie-wang/

Bunce, Susannah, 2019. *Sustainability, Planning and Gentrification of Cities.* London: Routledge.

Buntin, J., January 14, 2015. The Myth of Gentrification: It's Extremely Rare and Not as Bad for the Poor as You Think. *Slate.* http://www.slate.com/articles/news_and_politics/politics/2015/01/the_ gentrification_myth

Burns, Robyn and Lisbeth A. Berbary, 2020. Placemaking as Unmaking: Settler Colonialism, Gentrification, and the Myth of 'Revitalized' Urban Spaces. *Leisure Sciences: An Interdisciplinary Journal*: 1–8 (e-version). https://www.tandfonline.com/doi/pdf/10.1080/01490400.2020.1870592?needAccess=true

Buzar, Stefan, Philip Ogden, Ray Hall, A. Haase, S. Kabisch, and S. Steinführer, 2007a. Splintering Urban Populations: Emergent Landscapes of Reurbanisation in Four European Cities. *Urban Studies* 44(4): 651–677.

Buzar, Stefan, Ray Hall, Philip Ogden, 2007b. Beyond Gentrification: The Demographic Reurbanisation of Bologna. *Environmental and Planning A* 39: 64–85.

Byrne, Peter, J., 2003. Two Cheers for Gentrification. *Howard Law Journal* 46(3): 405–432.

Capps, Kriston, July 16, 2019. The Hidden Winners in Neighborhood Gentrification. Bloomberg CityLab. Bloomberg.com/news/articles/2019-07-16/the-hidden-winners-in-neighborhood-gentrification

Carlsten, Stephen, Gregory, 2019. Gentrification Without the Negative: A Rhetorical Analysis of the Franklin Neighborhood. Undergraduate Thesis. Columbus: The Ohio State University.

Casque, Francisco Diaz, 2013. Race, Space, and Contestation: Gentrification in San Francisco's Latina/o Mission District, 1998–2002. Doctoral Dissertation. University of California, Berkeley, CA.

Christophers, Brett, 2011. Revisiting the Urbanization of Capital. *Annals of the Association of American Geographers* 101(6): 1347–1364.

Cooper, Frederick, 1997. Modernizing Bureaucrats, Backward Africans, and the Development Concept. In Cooper, Frederick and R. Packard, eds. *International Development and the Social Sciences. Essays on the History and Politics of Knowledge.* Berkeley and Los Angeles: University of California Press: 64–92.

Cortright, Joe, March 10, 2020. Is Non-Gentrification the Real Threat to Neighborhoods? Rice Kinder Institute for Urban Research. Houston, TX: Rice University.

Cortright, Joe, July 19, 2019. How Gentrification Benefits Long-Time Residents of Low Income Neighborhoods. *City Commentary.* https://cityobservatory.org/how-gentrification-benefits-long-time-residents-of-low-income-neighborhoods/

Cortright, Joe, August 28, 2018. Do Rich Neighbors Make Low Income People Unhappy? *City Commentary.* https://cityobservatory.org/do-rich-neighbors-make-low-income-people-unhappy/

Couture, Victor and Jessie Handbury, 2015. Urban Revival in America, 2000 to 2010. Paper presented at the Research Symposium on Gentrification and Neighborhood Change, May 25, 2016. Philadelphia: Federal Reserve Bank of Philadelphia.

Clark, Julie, Ade Kearns, and Claire Cleland, 2016. Spatial Scale, Time and Process in Mega-Events: The Complexity of Host Community Perspectives on Neighborhood Change. *Cities* 53: 87–97.

Crush, Jonathan, 1995. Imagining Development. In Crush, ed. *Power of Development.* London: Routledge.

Danewid, Ida, 2020. The Fire This Time: Grenfell, Racial Capitalism and the Urbanization of Empire. *European Journal of International Relations* 26(1): 289–313.

Daniels, Ronald. April 5, 2019. Introductory Remarks. National Emergency Summit on Gentrification. C-SPAN.org. https://ibw21.org

Davidson, Justin, January 31, 2014. Is Gentrification All Bad? *New York* 47(3). https://nymag.com/news/features/gentrification-2014-2/

Davis, Keandra, February 3, 2020. Wicked Problems, Wicked Solutions: The Damaging Impacts of Gentrification on Urban Communities. Is Enough Being Done to Address This Issue? College of Social Sciences and Public Policy. Florida State University. https://wicked-solutions.blog/2020/02/03/the-damaging-impacts-of-gentrification-displacement-on-urban-communities-is-enough-being-done-to-address-this-issue/

Davis, Mike, 2007. Introduction. In Mike Davis and Daniel Bertrand, eds., *Evil Paradises: Dreamworlds of Neoliberalism.* New York: New Press.

DeVerteuil, Geoffrey, 2011. Evidence of Gentrification-Induced Displacement Among Social Services in London and Los Angeles. *Urban Studies* 48(8): 1563–1580.

Ding, Lei, Jackelyn Hwang, and Eileen Divringi, 2016. Gentrification and Residential Mobility in Philadelphia. Discussion Papers, Community Development Studies and Education Department. Philadelphia: Federal Reserve Bank of Philadelphia.

Du Bois, W.E.B., 1920. *Darkwater: Voices from Behind the Veil.* New York: Harcourt, Brace and Howe.

Eagleton-Pierce, Matthew, 2016. *Neoliberalism: The Key Concepts.* London: Routledge.

Edlund, Lena, Cecilia Machado, and Michaela Sviatchi, 2016. Bright Minds, Big Rent: Gentrification and the Rising Returns of Skill. Paper presented at the Research Symposium on Gentrification and Neighborhood Change, May 25, 2016. Philadelphia: Federal Reserve Bank of Philadelphia.

Ellen Gould, Ingrid, 2018. Can Gentrification Be Inclusive? In Herbert, Christopher, Jonathan Spader, Jennifer Molinsky, and Shannon Rieger, eds. *A Shared Future: Fostering Communities of Inclusion in an Era of Inequality.* Harvard: Joint Center of Housing Studies: 334–339.

Escobar, Arturo, 2000. Beyond the Search for a Paradigm? Post-Development and Beyond. *Development* 43(4): 11–14.

European American Chamber of Commerce, March 13, 2011. New York City Is Most Competitive City in the World. https://eaccny.com/news/new-york-city-is-most-competitive-city-in-the-world/

Field, Benjamin, 2016. Why Low-Income Housing Tax Credits Are Flowing to Gentrifying Neighborhoods. https://repository.library.georgetown.edu/bitstream/handle/10822/1059613/Shell_georgetown_0076M_14556.pdf?sequence=1&isAllowed=y

Freedman, Des, 2019. Media and the Neoliberal Swindle: From 'Fake News' to 'Public Service.' In Simon Dawes and Marc Lenormand, eds. *Neoliberalism in Context: Governance, Subjectivity and Knowledge.* Palgrave Macmillan: 215–231 (e-version).

Freeman, Lance, June 3, 2016. Five Myths about Gentrification. *The Washington Post.*

Freeman, Lance, 2005. Displacement or Succession? Residential Mobility in Gentrifying Neighborhoods. *Urban Affairs Review,* 40(4), 463–491.

Freeman, Lance and Frank Braconi, 2004. Gentrification and Displacement in New York in the 1990s. *Journal of the American Planning Association* 70(1): 39–52.

Freeman, Lance and Frank Braconi, 2002. Gentrification and Displacement. *The Urban Prospect* 1(2): [...].

Florida, Richard, 2017. *The New Urban Crisis: How Our Cities Are Increasing Inequality, Deepening Segregation, and Failing the Middle Class – And What We Can Do About It.* New York: Basic Books.

Florida, Richard, 2005. *Cities and the Creative Class: And How It's Transforming Work, Leisure, Community, and Everyday Life*. New York and London: Routledge.

Florida, Richard, 2002. *The Rise of the Creative Class*. New York: Basic Books.

Florida, Richard, January 30, 2017. More Losers Than Winners in America's New Economic Geography. *CityLab*. https://www.bloomberg.com/news/articles/2013-01-30/more-losers-than-winners-in-america-s-new-economic-geography

Florida, Richard and McLean, Jodie W., 2017. What Inclusive Urban Development Can Look Like. *Harvard Business Review*, July 11. https://hbr.org/search?search_type=&term=florida+and+Mclean

Ghertner, Asher, D., 2015. Why Gentrification Theory Fails in 'Much of the World.' *City* 19(4): 552–563.

Ghertner, Asher, D., 2014. India's Urban Revolution: Geographies of Displacement Beyond Gentrification. *Environment and Planning A* 46: 1554–1571.

Gibson, D.W., 2015. *The Edge Becomes the Center: An Oral History of Gentrification in the 21st Century*. New York: Overlook Press.

Gill, Stephen, 2003. *Power and Resistance in the New World Order*. New York: Palgrave Macmillan.

Gilroy, Paul, 2000. *Against Race: Imagining Political Culture Beyond the Color Line*. Cambridge, MA: The Belknap Press of Harvard University Press.

Godsil, Rachel, May 2014. Transforming Gentrification into Integration. The Dream Revisited: Advancing Research and Debate on Housing, Neighborhoods, and Urban Policy. NYU Furman Center. https://furmancenter.org/research/iri/essay/transforming-gentrification-into-integration

Goo, Delia Ah, 2018. Gentrification in South Africa: The 'Forgotten Voices' of the Displaced in the Inner City of Johannesburg. In Julie Clark and Nicholas Wise, eds. *Urban Renewal, Community and Participation: Theory, Policy and Practice*. Springer: 89–100.

Glass, Ruth, L., 1964. *London: Aspects of Change*. Edited by Center for Urban Studies. London: MacGibbon & Kee.

Glazer, Lou, September 6, 2019. Positives of Gentrification Outweigh the Negatives. *Grand Rapids Business Journal*. https://grbj.com/opinion/positives-of-gentrification-outweigh-the-negatives/

Gotham, Kevin, Fox, 2001. Redevelopment for Whom and for What Purpose? A Research Agenda for Urban Redevelopment in the Twenty First Century. In Gotham, Kevin, Fox, ed. *Critical Perspectives on Urban Redevelopment:* Volume 6 of *Research in Urban Sociology 6*. New York: Emerald Press.

Hakimi, David, P., March 23, 2019. Gentrification Done Right Can Bridge the Class Divide. Language of Light blogpost. https://www.alconlighting.com/blog/learning-lab/gentrification-done-right-can-bridge-class-divide/

Hamnett, Chris, 1991. The Blind Men and the Elephant: The Explanation of Gentrification. *Transactions of the Institute of British Geographers* 16(2): 173–189.

Hartley, Daniel, A., Nikhil Kaza, and T. William Lester, 2016. Are America's Inner Cities Competitive? Evidence From the 2000s. *Economic Development Quarterly* 30(2): 137–158.

Harvey, David, 2017. *Marx, Capital and the Madness of Economic Reason*. Oxford: Profile Books.

Harvey, David, 2012. *Rebel Cities: From the Right to the City to the Urban Revolution*. New York: Verso.

Harvey, David, 2009. The Crisis, Our Challenge. Interview with Marco Berlinguer and Hilary Wainwright. http://www.redpepper.org.uk/Their-crisis-our-challenge.

Harvey, David, 2008. The Right to the City. *New Left Review* 53: 23–40.

Harvey, David, 2007. Neoliberalism as Creative Destruction. *The Annals of the American Academy* 610(1): 21–44.

Harvey, David, 2007a. *A Brief History of Neoliberalism.* New York: Oxford University Press.

Harvey, David, 2003. *The New Imperialism.* Oxford: Oxford University Press.

Harvey, David, 2003a. The Right to the City. *International Journal of Urban and Regional Research* 27(4): 939–941.

Harvey, David, 1985. *The Urbanization of Capital: Studies in the History and Theory of Capitalist Urbanization.* Oxford, UK: Blackwell.

Harvey, David and David Wachsmuth, 2012. "What Is to Be Done? And Who the Hell Is Going to Do It?" In Brenner, Neil, Peter Marcuse, and M. Mayer, eds. *Cities for People, Not for Profit: Critical Urban Theory and the Right to the City.* London: Routledge: 264–274.

Haymes, Stephen, Nathan, 1995. *Race, Culture, and the City: A Pedagogy for Black Urban Struggle.* New York: State University of New York Press.

Herscher, Andrew, 2020. The Urbanism of Racial Capitalism: Towards a History of 'Blight.' *Comparative Studies in South Asia, African and the Middle East* 40(1): 57–65.

Hill Collins, Patricia, 2002. Between Camps/Against Race. *Ethnicities* 2(4): 539–544.

Hollands, Robert and Paul Chatterton, 2003. Producing Nightlife in the New Urban Entertainment Economy: Corporatization, Branding and Market Segmentation. *International Journal of Urban and Regional Research* 27(2): 361–385.

Holm, Andrej, 2013. Berlin's Gentrification Mainstream. In Britta Grell, Mathias Bernt, and Andrej Holms, eds. *The Berlin Reader: A Compendium on Urban Change and Activism.* London: Verlag: 171–187.

Hopkinson, Natalie, September 9, 2012. Gentrification: Views from Both Sides of the Street/Farewell to Chocolate City. *New York Times.*

Howland, Steven, Anthony, 2020. I Should Have Moved Somewhere Else: The Impacts of Gentrification on Transportation and Social Support for Black Working-Class Families in Portland, Oregon. Doctoral Dissertation: Portland State University.

Hymowitz, Kay, S., November 13, 2015. Op-Ed: The Mistaken Racial Theory of Gentrification. *Los Angeles Times.*

Hudson, Michael, 2020. Rentiers a Bunch of Gangsters. Michael Hudson: On Finance, Real Estate and the Powers of Neoliberalism. https://michael-hudson.com/2021/01/rentiers-a-bunch-of-gangsters/

Issar, Siddhant, 2020. Listening to Black Lives Matter: Racial Capitalism and the Critique of Neoliberalism. *Contemporary Political Theory* 20(1): 8–24 (e-version).

Jackson, Liza Kim, 2017. The Complications of Colonialism for Gentrification Theory and Marxist Geography. *Journal of Law and Social Policy* 27: 43–71.

Janoschka, Michael and Jorge Sequera, 2016. Gentrification in Latin America: Addressing the Politics and Geographies of Displacement. *Urban Geography* 37(-8): 1175–1194.

Kasdin, Neisen, February 25, 2020. Opinion: Why Gentrification is Good. *Miami Herald.* https://www.miamiherald.com/news/business/real-estate-news/article240612146.html

Katz, Bruce and Julie Wagner, 2014. *The Rise of Innovation Districts: A New Geography of Innovation in America.* Washington, D.C: Brookings Institute, Metropolitan Policy Program. https://c24215cec6c97b637db6-9c0895f07c3474f6636f95b6bf3db172.ssl. cf1.rackcdn.com/content/metro-innovation-districts/~/media/programs/metro/ images/innovation/innovationdistricts1.pdf

J.T. The LA Storyteller, December 2020. It's Now Time to Ask if Gentrification is a Fourth Wave of Jim Crow Policy. Knock LA. https://www.chicano.ucla.edu/files/ news/It%E2%80%99s%20Now%20Time%20to%20Ask%20if%20Gentrification %20is%20a%20Fourth%20Wave%20of%20Jim%20Crow%20Policy%20by%20 J.T.%20The%20L.A.%20Storytelle_KNOCK_121620.pdf

Layser, Michelle, D., 2019. The Pro-Gentrification Origins of Place-Based Investment Tax Incentives and a Path Towards Community Oriented Reform. Unpublished manuscript. https://lowellmilkeninstitute.law.ucla.edu/wp-content/uploads/2019/01/ Layser-The-Pro-Gentrification-Origins-WIP.pdf

Lees, Loretta, 2011. The Geography of Gentrification: Thinking Through Comparative Urbanism. *Progress in Human Geography* 36(2): 155–171.

Lees, Loretta, 2003. Super-Gentrification: The Case of Brooklyn Heights, New York City. *Urban Studies* 40(12): 2487–2509.

Lees, Loretta, 2000. A Reappraisal of Gentrification: Towards a Geography of Gentrification. *Progress in Human Geography* 24: 389–408.

Lees, Loretta, Hyun Bang Shin, and Ernesto López-Morales, 2016. *Planetary Gentrification.* London: Wiley.

Lees, Loretta, Tom Slater, and E. Wyler, 2008. *Gentrification.* New York: Routledge.

Lefebvre, Henri, 2003. *The Urban Revolution.* Minneapolis: University of Minnesota Press.

Lefebvre, Henri, 1991. *The Production of Space.* Oxford: Blackwell.

Lewis, Nemoy, 2020. Anti-Blackness Beyond the State: Real Estate Finance and the Making of Urban Racial Capitalism. *State and Space.* https://www.societyand-space.org/articles/anti-blackness-beyond-the-state-real-estate-finance-and-the-making-of-urban-racial-capitalism

Ley, David, 1980. Liberal Ideology and the Postindustrial City. *Annals of the Association of American Geographers* 70: 238–258.

Ley, David, 1996. *The New Middle Class and the Remaking of the Central City.* Oxford: Oxford University Press.

Lipman, Pauline, 2018. Segregation, the 'Black Spatial Imagination,' and Radical Social Transformation. *Democracy & Education* 26(2): 1–8.

Lipsitz, George, 2018. Living Downstream: The Fair Housing Act at Fifty. In Gregory D. Squires, ed. *The Fight for Fair Housing: Causes, Consequences and Future Implications of the 1968 Federal Fair Housing Act.* New York: Routledge: 266–290.

Lipsitz, George, 2011. *How Racism Takes Place.* Philadelphia: Temple University Press.

Lipsitz, George, 2006 [1998]. *Possessive Investment in Whiteness: How White People Profit from Identity Politics.* (revised and expanded edition) Philadelphia, PA: Temple University Press.

Lovering, John, 2007. The Relationship Between Urban Regeneration and Neoliberalism: Two Presumptuous Theories and a Research Agenda. *International Planning Studies* 12(4): 343–366.

Madden, David, October 10, 2013. Gentrification Doesn't Trickle Down to Help Everyone. *The Guardian.*

Martin, Michel, March 13, 2014. Gentrification: Progress or Destruction? *NPR.* https://www.npr.org/2014/03/13/289798823/gentrification-progress-or-destruction

Martinot, Steve, 2003. *The Rule of Racialization, Class Identity, Governance.* Philadelphia, PA: Temple University Press.

Marx, Karl, 1964. *Karl Marx: Selected Writings in Sociology & Social Philosophy.* New York: McGraw-Hill.

Matthews, Chris, July 29, 2016. Debunking the Gentrification Myth. Fortune. https://fortune.com/2016/07/29/gentrification-housing-prices/

McCabe, Brian, J., 2019. Protecting Neighborhoods or Priming Them for Gentrification? Historic Preservation, Housing, and Neighborhood Change. *Housing Policy Debate* 29(1): 181–183.

McElroy, Erin and Werth, Alex, 2019. Deracinated Dispossessions: On the Foreclosures of 'Gentrification' in Oakland, CA. *Antipode* 51(3): 878–898.

McMillan Cottom, Tressie, 2020. Where Platform Capitalism and Racial Capitalism Meet: The Sociology of Race and Racism in the Digital Society. *Sociology of Race and Ethnicity* 6(4): 441–449.

Mendes, Luís, 2018. Gentrification and the New Urban Social Movements in Times of Post-Capitalist Crisis and Austerity Urbanism in Portugal. *Arizona Journal of Hispanic Cultural Studies* 22: 199–125.

Mills, Charles, M., 1997. *The Racial Contract.* Ithaca, NY: Cornell University Press.

Millsap, Adam, A., March 29, 2018. We Shouldn't Stop Gentrification, But We Can Make It Less Painful. *Forbes.* https://www.forbes.com/sites/adammillsap/2018/03/29/we-shouldnt-stop-gentrification-but-we-can-make-it-less-painful/?sh=26be353a7ba0

Mirowski, Philip and Dieter Plehwe, 2009. *The Road from Mont Pèlerin: The Making of the Neoliberal Thought Collective.* Cambridge, MA: Harvard University Press.

Misoczky, Maria, Ceci and Clarice Misoczky de Oliveira, 2018. The City and the Urban as Spaces of Capital and Social Struggle: Notes on Henri Lefebvre's Enduring Contributions. *Brazilian Journal of Public Administration* 52(6): 1015–1031.

Mitchell, Shekinah, 2019. *In Richmond, Virginia, Gentrification Is Colonization.* Washington, DC: National Community Reinvestment Coalition (NCRC). See https://ncrc.org/gentrification-richmondva/

Moss, Jeremiah, 2018. Hyper-Gentrification in the Revanchist City. RNA History Club Session 25, December 9. http://www.columbia.edu/~hauben/RNA-House/history/session25-text.pdf

Moss, Jeremiah, 2017. *Vanishing New York: How a Great City Lost Its Soul.* New York: Dey Street Books.

Müller, Martin, 2015. The Mega-Event Syndrome: Why So Much Goes Wrong in Mega-Event Planning and What to Do About It. *Journal of the American Planning Association* 81(1): 6–17.

Mumm, Jesse, 2017. The Racial Fix: White Currency in the Gentrification of Black and Latino Chicago. *Focaal-Journal of Global and Historical Anthropology* 79: 102–118.

Mumm, Jesse Stewart, 2014. When the White People Come: Gentrification and Race in Puerto Rican Chicago. Doctoral Dissertation. Evanston, Illinois: Northwestern University.

Naram, Kartik, 2017. No Place Like Home: Racial Capitalism, Gentrification, and the Identity of Chinatown. *Asian American Policy Review* 26: 1–26.

Nelson, Glenn, November 21, 2019. How Seattle Can Slow Gentrification – And Why It Must Try. *Crosscut*. https://crosscut.com/opinion/2019/11/how-seattle-can-slow-gentrification-and-why-it-must

Nevius, C.W., 2013. Gentrification No Longer a Dirty Word. *San Francisco Chronicle*.

Newman, Kathe and Elvin, K. Wyly, 2006. The Right to Stay Put, Revisited: Gentrification and Resistance to Displacement in New York City. *Urban Studies* 43(1): 23–57.

Ortega, Arnisson Andre, C., 2016. Manila's Metropolitan Landscape of Gentrification: Global Urban Development, Accumulation by Dispossession & Neoliberal Warfare Against Informality. *Geoforum* 70: 35–50.

Patillo, Mary, E., 2007. *Black on the Block: The Politics of Race and Class in the City.* Chicago: University of Chicago Press.

Perry, Andre, December 7, 2018. Homeowners Have Lost $156 Billion by Living in a 'Black Neighborhood.' *CNN Business*. https://www.cnn.com/2018/12/06/perspectives/black-home-ownership-undervalued-brookings/index.html

Pieterse, Jan Nederveen, 2010. *Development Theory: Deconstructions/Reconstructions.* New York: Sage Publications.

Pieterse, Jan Nederveen, 2000. After Post-Development. *Third World Quarterly* 25: 231–254.

Povinelli, Elizabeth, 2011. *Economics of Abandonment.* Durham, NC: Duke University Press.

Powell, John, A. and Marguerite L. Spencer, 2003. Giving Them the Old 'One-Two': Gentrification and the K.O. of Impoverished Urban Dwellers of Color. *Howard Law Journal* 46(3): 433–491.

Quizar, Jessi, 2020. Comeback as Re-Settlement: Detroit, Anti-Blackness, and Settler Colonialism. *Items*. Social Science Research Council. https://items.ssrc.org/layered-metropolis/comeback-as-re-settlement-detroit-anti-blackness-and-settler-colonialism/

Quizar, Jessi, 2019. Land of Opportunity: Anti-Black and Settler Logics in the Gentrification of Detroit. *American Indian Culture and Research Journal* 43(2): 113–133.

Ramirez, Margaret, M., 2020. City as Borderland: Gentrification and the Policing of Black and Latinx Geographies in Oakland. *EPD: Society and Space* 38(1): 147–166.

Redmon, Kevin, Charles, May 18, 2010. The Man Who Reinvented the City. *The Atlantic*.

Rice, Jennifer, L., Daniel Aldana Cohen, Joshua Long, and Jason R. Jurjevich, 2019. Contradictions of the Climate-Friendly City: New Perspectives on Eco-Gentrification and Housing Justice. *International Journal of Urban and Regional Research* 44(1): 145–165.

Right to the City Montreal, 2012. Colonizing the Inner-City – Gentrification and the Geographies of Colonialism. https://rightothecitymtl.wordpress.com

Robinson, Cedric, 2000 [1983]. *Black Marxism: The Making of the Black Radical Tradition.* 2nd ed. Chapel Hill: University of North Carolina.

Romano, Andrew and Garance Franke-Ruta, March 5, 2018. A New Generation of Anti-Gentrification Radicals Are on the March in Los Angeles and Around the Country. *Huffington Post*.

Roy, Ananya, March 5, 2018. Racial Banishment: A Postcolonial Critique of the Urban Condition in America. Presentation at Unit for Criticism and Interpretative Theory, University of California, L.A.

Sanneh, Kelefa, July 4, 2016. Is Gentrification Really a Problem? *The New Yorker.*

Shell, Jeanna, 2020. The Low-Income Housing Tax Credit: Affordable Housing Producer or Gentrification Generator? Doctoral Dissertation. Georgetown University. https://repository.library.georgetown.edu/handle/10822/1059613

Sen, Amartya, K., 1999. *Development as Freedom.* New York: Anchor Books.

Sevilla-Buitrago, Alvaro, 2013. Debating Contemporary Urban Conflicts: A Survey of Selected Authors. *Cities* 31: 454–468.

Shaw, Joe, 2020. Hyperreal Estate: The Production of New Urban Real Estate Markets. Doctoral Dissertation. Oxford: Oxford University. https://ethos.bl.uk/OrderDetails.do?uin=uk.bl.ethos.823576

Shin, Byun Bang, Lees Loretta, and Lopez-Morales, Ernesto, 2015. Introduction: Locating Gentrification in the Global East. *Urban Studies* 53(3): 455–470.

Slater, Tom, 2017. Planetary Rent Gaps. *Antipode* 49(S1): 114–137.

Slater, Tom, 2014. Unravelling False Equivalence Urbanism. *Analysis of Urban Change, Theory, Action* 18(4–5): 517–524.

Slater, Tom, 2012. Missing Marcuse: On Gentrification and Displacement. In Brenner, Neil, Peter Marcuse, and M. Mayer, eds. *Cities for People Not for Profit: Critical Urban Theory and the Right to the City.* New York: Routledge: 171–196.

Slater, Tom, 2011. Chapter 50: Gentrification of the City. In Bridge, Gary and Sophie Watson, eds. *The New Blackwell Companion to the City.* New York: Blackwell Publishing Company: 571–585.

Slater, Tom, 2006. The Eviction of Critical Perspectives from Gentrification Research. *International Journal of Urban and Regional Research* 30(4): 737–757.

Slobodian, Quinn, 2018. *Globalists: The End of Empire and the Birth of Neoliberalism.* Cambridge: Harvard University Press.

Smart, A. and Smart, J., 2017. Ain't Talkin' 'Bout Gentrification: The Erasure of Alternative Idioms of Displacement Resulting from Anglo-American Academic Hegemony. *International Journal of Urban and Regional Research* 41(3): 518–525.

Smith, Neil, 2002. New Globalism, New Urbanism: Gentrification as Global Urban Strategy. *Antipode* 427–450.

Smith, Neil, 2001. Global Social Cleansing: Postliberal Revanchism and the Export of Zero Tolerance. *Social Justice* 28(3): 68–74.

Smith, Neil, 1996. *The New Urban Frontier: Gentrification and the Revanchist City.* London: Routledge.

Smith, Neil, 1992. Blind Man's Bluff, or Hamnett's Philosophical Individualism in Search of Gentrification. *Transactions of the Institute of British Geographers* 17(1): 110–115.

Smith, Neil, 1984. *Uneven Development: Nature, Capital, and the Production of Space.* New York: Basil Blackwell.

Smith, Neil, 1979. Towards a Theory of Gentrification: A Back to the City Movement by Capital, Not People. *Journal of the American Planning Association* 45(4): 538–548.

Smith, Neil and James deFilippis, 1999. The Reassertion of Economics: 1990s Gentrification in the Lower East Side. *International Journal of Urban and Regional Research* 23(4): 638–653.

Smith, R. Thomas, 1919. Charles Baudelaire: The Generous Player. In Thomas Robert Smith, ed. *Smith. Baudelaire: His Prose and Poetry.* New York: Modern Library: 82.

Spinney, Justin, 2021. *Understanding Urban Cycling: Exploring the Relationship Between Mobility, Sustainability and Capital.* London: Routledge.

Star Tribune Editorial Board, August 19, 2015. A Little Gentrification Can Be a Good Thing for North Minneapolis. *Star Tribune*. https://www.startribune.com/-a-little-gentrification-can-be-a-good-thing-for-north-minneapolis/322340901/

Stein, Samuel, 2019. *Capital City: Gentrification and the Real Estate State*. New York: Verso.

Stovall, David, April 12, 2021. Black Urban Crises and Push-Out 'Engineered.' *Black Agenda Report*. https://soundcloud.com/user-92939733/black-urban-crises-and-push-out-engineered

Stovall, David, November 21, 2016. Engineered Conflict. Public Lecture. University of Illinois at Chicago (UIC) Great Cities Institute, Chicago. https://www.youtube.com/watch?v=F7BeOUXMSro

Su, Yichen, 2019. *Gentrification Transforming Neighborhoods in Big Texas Cities*. Dallas: Federal Reserve Bank of Texas.

Sullivan, L., 2014. Gentrification May Actually Be Boon to Longtime Residents. *NPR. org*. http://www.npr.org/2014/01/22/264528139/long-a-dirty-word-gentrification-may-actually-be-losing-its-stigma

Sutton, Stacey, 2020. Gentrification and the Increasing Significance of Racial Transition in New York City, 1970–2010. *Urban Affairs Review* 56(1): 65–95.

Swanson, Kate, 2007. Revanchism Heads South: The Regulation of Indigenous Beggars and Street Vendors in Ecuador. *Antipode* 39(4): 708–728.

The Economist, June 21, 2018. In Praise of Gentrification. https://www.economist.com/united-states/2018/06/21/in-praise-of-gentrification

Urena-Ravelo, Briana, L., 2017. It's True, Gentrification Isn't the New Colonialism, It's Just the Old One. *Medium*. https://medium.com/@AfroResistencia/its-true-gentrification-isn-t-the-new-colonialism-it-s-just-the-old-one-daf7e97a86f0

Van den Bosch, July 14, 2017. How Google Connects with the Smart City Movement. *Smart City Hub*. https://smartcityhub.com/technology-innnovation/google-connects-smart-city-movement/

Van der Hoeven, Arno and Erik Hitters, 2020. The Spatial Value of Live Music: Performing (Re)Developing and Narrating Urban Spaces. *Geoforum* 117: 154–164.

Vigdor, Jacob, 2010. Is Urban Decay Bad? Is Urban Revitalization Bad Too? *Journal of Urban Economics* 68: 277–289.

Vigdor, Jacob L., Douglas S. Massey, and Alice M. Rivlin, 2002. Does Gentrification Harm the Poor? *Brookings-Wharton Papers on Urban Affairs*: 133–182.

Wacquant, Loïc, 2008. Relocating Gentrification: The Working Class, Science and the State in Recent Urban Research. *International Journal of Urban and Regional Research* 23(1): 198–205.

Wagner, Jacob, 2017. 'Negro Removal' Revisited: Urban Planning and the New Jim Crow in Kansas City. *Progressive City*. https://www.progressivecity.net/single-post/2017/03/12/-negro-removal-revisited-urban-planning-and-the-new-jim-crow-in-kansas-city

Wahba, Dina, 2020. Urban Rights and Local Politics in Egypt: The Case of the Maspero Triangle. Arab Reform Initiative: Research Papers.

Wang, Jackie, 2018. *Carceral Capitalism*. Semiotext(e)/Intervention Series 21. Cambridge, MA: MIT Press.

Wharton, Jonathan, L., 2008. Gentrification: The New Colonialism in the Modern Era. *Forum on Public Policy: A Journal of the Oxford Round Table*. https://books.apple.com/us/book/gentrification-the-new-colonialism-in-the-modern-era/id512794633

Wilson, David, 2004. Towards a Contingent Urban Neoliberalism. *Urban Geography* 25(8): 771–783.

Wilson, William, Julius, 1987. *The Truly Disadvantaged: The Inner City, the Underclass, and Public Policy*. Chicago, IL: University of Chicago Press.

World Bank, no date. Managing the Potential Undesirable Impacts of Urban Regeneration: Gentrification and the Loss of Social Capital. https://urban-regeneration. worldbank.org/node/45

Wyly, Elvin, K., 2015. Gentrification on the Planetary Frontier: The Evolution of Turner's Noösphere. *Urban Studies* 52(14): 2515–2550.

Wynn, Jonathan and Andrew Deener, October 10, 2017. Gentrification? Bring It On. *The Conversation*. https://theconversation.com/gentrification-bring-it-82107

Wynne, Derek, 1998. Book Review of the New Middle Class and the Remaking of the Central City by David Ley. *Work, Employment & Society* 12(1): 182–184.

Zimmer, Tyler, J., 2020. Gentrification and the Racialization of Space. *Philosophy and Social Criticism*: 1–21.

Zukin, Sharon, 2010. *Naked City: The Death and Life of Authentic Urban Places*. Oxford: Oxford University Press.

Interviews from the Field

9 How CRED Financing Differs from Market Rate?

Joe Bonora

Q: How would you on the finance side, just out of the gate, structure your performance differently? Go about it differently than you would maybe on a pure market rate.

R: So, when you're saying like community development, you're saying a project that has components within it that benefit the overall community, not just the developer in terms of economics, right?

Q: Community is at the core of CRED. One of the things we talk about in CRED is that we considered roughly 20% of a CRED proforma will be community-driven. How would you go about doing a CRED type of proforma versus a market rate proforma?

R: So, all of our projects, typically we try to incorporate into them some type of community benefit. So, whether it's developing a project in an area that would be considered blighted or underdeveloped or underutilized land, so that could be part of the community development component of it. But as it relates to like setting aside units for workforce or affordable housing, or included within the project, like a community center, Like a boys and Girls Club or YMCA or something like that. Financing it. You really try to break out the cost of that. So, for example, for housing, if you were going to allocate and set aside 20% of units for, let's just say workforce because I don't want to because like affordable housing being 60% AMI or less. Which is eligible for tax credits. That's a whole different conversation. That's a different type of financing. But let's say you wanted to set aside units for workforce housing, which would be 80–120% AMI, we would determine like in doing that, what are the rents that we have to? What would be Captiva rents at to accommodate for that? And then what's that delta between those rents and market rate rents? And then that total amount that difference, we basically capitalized that figure out what that value differences, and then that's how we started looking at getting incorporating incentives, economic development incentives. So, structurally what we do is really trying to upfront quantify the cost of those community benefits and then figure out how to pull in additional funding to offset that to make the economics of the deal work.

DOI: 10.1201/9781003109679-11

Q: When you say quantify, I'm looking at this from two perspectives, you're of course quantifying it in your own development. To offset those lost to non-market rate, you may need to jack up the market rate side or get tax credits. But when you are talking with the community, it becomes a bit vague. How do you quantify that?

R: Generally, again we look at doing an impact study that includes everything from building a project in the area and introducing 300 new units of residential housing and how does that affect the area, meaning the local spending. So, now you've got 300 more residents or 300 plus more residents spending money at the retail locations around it. You've got a base of people that could be employees for firms that would then incentivize and induce a company to move down there as well, so bringing jobs, the area and so there are impacts that come off of that and so we do try to sometimes that is easy to quantify that because then you start getting into like these different offshoots of like the impacts as it goes out. But we do try to figure out how much of what that benefit is. Try to assign a number to it if possible, that how much of the benefit the municipality in their communities get in and then also qualify it and show what exactly those benefits are beyond just the economics of it. We're trying to show both sides of it.

Q: So then when you're going to the city to or municipalities substructure to ask for incentives, how do you say? How do you spin that and what are the most common kind of incentives you might be able to get your projects done?

R: Spin maybe puts into a negative light. So, let's just say, how do you best frame your request? And how do you show a group whether it's commissioners or just the public who may not have background in finance or really understand some of these financial impacts. How do you basically make it simple to understand? And so we've done a couple of different ways. The first time we did it, tax increment financing request with the CRA in Fort Myers. We essentially had a show that without the incentive, if we built this particular project on this site, the rents that we would get would not generate an IRR that would meet the requirements of most investors in the marketplace. An investor would typically want to see an unlevered return of a number maybe 10, it could be 12, or could be 20%. And then we would show that without the incentive this project really generates 8% and so in order to hit that return requirement we need this amount of incentive. So, that's just a different kind of cash flow model. So, we'd layer into the model, like here is it without the incentive and here's it with the incentive, which is a rebate of the ad valorem taxes. So, we've done that. Then, what we've done more recently is we looked at different projects we're building a project in the urban core of downtown and what's the difference between that project and a project that's being built out by the highway the commissioners, and elected officials are saying, they're getting pushed back because all these other apartment units are getting built without incentives. There are three main items. Obviously parking garage, the costs of structured parking versus the costs of doing a surface park asphalt, there's a significant difference there. Then, there is stormwater and doing a vault. Could dig a retention or detention pond and then finally just having the ability to have enough land to stage, have parking and equipment and all that there when you don't have all of that in your downtown. Working within this little site, the costs go up. The cost of trying to manage the crane and the parking and leasing extra land so you can store equipment. So, we came up with the total number to say that in order to build downtown, it's going to cost 6 million, 7 million, 10 million more than it would cost to build the same project out in the suburbs right out an area that's got 25 acres plus. And so that's how we'd go about asking for that number. And let's just say it was $8 million. But we're getting an ad valorem tax rebate over 20 years. We would do a net present value. Meaning it's $8 million today to build this parking garage. We're going to get $15 million in 20 years. But obviously, the value of that stream of cash flow is worthless now than it is over time. So, we discount that back. We apply a reasonable rate 4 or 5%. We determine what the net present value is, and that's how we go about requesting 90% or 80% or whatever the percentages of ad valorem tax rebate annually for a set amount of time.

Q: When you do your capital stack for your projects, what are sources that you might use as a for profit versus a non-profit?

R: So as a for profit, which is the majority of our development is for profit, a typical capital structure for us on a project would be 65–70% of the total project cost. As senior debt being our first mortgage allow and either from a bank or a non-bank lender or finance company, or a debt fund or government agency like HUD. So, really the options today are bank or debt fund and a debt fund, pool of capital that you know essentially like a bank, but without all the regulation and limitations that banks have and each has their own pros and cons, we usually layer and subordinate to that would be either preferred equity or mezzanine financing, essentially the same, but structurally different. In that mezzanine financing is debt. Preferred equity is an equity layer that essentially is junior to the senior debt, but without any rights as a creditor. They both have similar cost structures that would generally represent about 15% of the total cost. And then the balance of it would be common equity. So, that's the developer and any investors at the developer brings in. So like us will invest in that level as the developer and the sponsor will invest as common equity and then we'll bring in a handful of outside investors on that level. That is a very basic capital structure, I think, used by most developers these days. Some will be elected, not use preferred equity or mez if they have a substantial amount of liquidity and they don't need it because they don't want to pay 9, 10% when they have cash on hand, but generally you're not going to find many developers that don't like to use leverage to the maximum extent possible. Again, not debt, so you really have about 30% equity in 70% debt.

On the non-profit side and non-profit development which is going to be more affordable housing. Although there are some ways of utilizing a non-profit to finance a project outside of tax credits. But I would say that most non-profit in residential housing is going to be for affordable tax credit. So LIHTC tax credit and so that capital structure is significantly more complicated. You have to set up a for profit entity. That would be the partnership that would own the property. At 99.9% of that entity will be owned by a tax credit investor. The non-profit would be the general partner in a limited partnership and they would come 0.1% of the entity.

We usually do not invest any of its own capital, LIHTC or affordable housing deals. The appeal of those to developers, both for and non-profit developers is the limited amount of upfront capital or investment capital you need. That's not to say you don't need money to do it, so generally on a normal-sized project of 100 and 120 units, developer would have to have between 800,000 and probably a million and a half dollars of cash into the deal that they would get reimbursed once they close. So, the net once you close in your financing is that you will have zero money into it but leading up to that point you do have to have cash to invest in these projects to get through the pursuit side of it. Pay the architect, pay everybody. Then generally when that's done, the transaction is structured through a combination of low-income housing tax credits, whether 9% or 4% the competitive or non-competitive tax credits. Bond financing if it's 4% or debt financing if it's a 9% deal and then on top of that if you have a 4% non-competitive deal, you're going to have the alphabet soup of other funding in there. That's CDBG, Home, etc. We had four or five different sources of funding you're trying to get as much as you can to fill that gap, and again, much more complicated because everybody has requirements attached to their funding and you got to make sure they all are in compliance with one another, so it's a lot trickier of a deal. So, market rate is a lot more streamlined and efficient to finance. Affordable housing becomes a lot more complicated and complex, which is probably, in my opinion, one of the reasons that we have a shortage of affordable housing.

Q: That was going to be my follow-up question. Right there you just answered, it was. Do you think the regulatory structure is really holding people back from doing it? Because on paper it looks like free money.

R: I'm one hundred percent confident that's the case because that's why we've decided to shy away from it. So, we've decided as a non-profit to be more on the finance side of those projects as opposed to the development side. If you were going to be an affordable housing developer, you have to be an affordable housing developer, meaning you're all in. This is what you do. It's not like we're going to do a one off affordable deal while we also do market, it's at some point in the future an affordable housing developer who's been doing it for a number of years and has a pipeline. And they've done a number of transactions. If they want to get into doing market rate, they probably can, but for market rate developer to dabble in affordable housing it's very difficult because it's a system itself. At least in Florida, it's kind of a perverse incentive system. They've created this competitive process that discourages collaboration and partnership between organizations trying to get affordable housing built. It's like the scoring model that if you win a competitive award, there is an incentive for other developers to contest it and try to get that award because they lost that because there are only so many credits to go around. And if there was a way to fix that system or come up with other ways to finance housing beyond just competitive credits, you'd see a lot more of it get built, but it is complicated. It's ultra-competitive and if you're an affordable housing developer or if you're considering doing either affordable market rate, if you were a new development and you just looked at the two transactions and what goes into each one of them both in time, process, thought, effort, I would say that 95 plus percent of the time you're going to do market rate.

It's just the path of least resistance. The economics are more attractive and you just have fewer limitations on what you could do. I mean, as a developer, my preference is just because I enjoyed the overall design and coming up with a plan and building something unique, a lot of that has to go away when you're getting into affordable because you're kind of working within the confines of the program and the design, and just making the numbers work. The complexity is a big thing. I think that's inhibiting the ability for people to do more of it.

Q: So, in today's market, you talked a little bit about mixing in the community benefit into your projects. Can you talk to us a little bit about mixed-use project where you might have residential units, and they're asking you to put in some commercial also?

I think what gets lost in a lot of developers, not because they don't want to do it, I think a lot of times in any business, whether it's private or even public sector, people tend to do what is easiest. It's the thing that if it requires more time or thought, generally speaking, people are going to go the path that's going to be a little bit less rocky. But there are things that you do while complicated upfront, enhance the value of a project overall while also enhancing the community. So, as an example, that we're doing a project now that we are, it's a mixed-use project. It's going to include some office space, will have a little bit of retail. We will essentially give away the retail space for free because the users of that space be a coffee shop, pet groomer, or all amenities for our residents that live there come in my projections when I'm building out a financial model for project. I'm not even using those rents the retail rents as part of my numbers. So I'm not even taking in consideration that income. So, really anything that you get is for your upside anyway. But you're not considering as part of your underwriting. I'd rather get somebody into that space. They could build it out themselves. They'll operate it, but they don't pay any rent. But now I have on site a coffee shop, I have on site a groomer, on site whatever. So, you try to integrate into that you can help a local small business getting into that space. You also enhance the community for your residence and then on the office side of it. We've been spending a lot of time on this project that we're working on now and trying to get a local organization that has a number of locations to consolidate and bring their headquarters to this site. And so, we spent a lot of time working with them, working with the CRA, and trying to come up with something that works for them. And that will make the economics, at least for us, worthwhile doing it. But, we could find another use for this parcel and make more money, but I think having those jobs and bring into this area is going to help stimulate everything else around the area. So, we've got a mixed-use project. It's the first project in a redevelopment area by bringing in 60,000 square feet of commercial space and having a user bring their headquarters there. That's going to bring 100 plus new workers. That's going to bring in and attract more companies to come down there. It's going to attract more residential, so it's going to help build on itself. And then as that area develops around it, it enhances the value of the project. So while on the one hand, it might be more time and you might not make as much money doing it upfront in the short term, in the long term, you're going to create a significant amount of value for yourself because you're going to enhance the community, and that's kind of a good example of how spending more time and trying to build out the community around the project benefit you, but you got to be in for the long term. That's the difference between getting in stabilizing or flipping, or project versus having a ten year plus view. But the value in real estate is all long term. I mean real estate is a long game if you go into it with that perspective that ten year plus view and you do things that enhance the area around the project the multiple that you make on your investment is going to be significantly higher than what you would make just getting it again now.

Q: That was really interesting, because it's one of these things I try to tell my students and sometimes you're going to take a little bit of a step back to take those two steps forward. Look, I like that point you're going to give you the free or really reduced rent to a groomer, whatever that is also amenities for building, but build up the whole community that ten years from now when we sell the building. So that was really kind of an interesting point you make there for sure.

R: I think that the difference between somebody would be considered probably more community development focus versus just regular market rate development is the investment horizon and the view that you're taking because I think it's clearly there's an economic case for doing all these things, but the time it takes to realize it is longer than most people want it. An investment fund, for example, has a ten-year window. But it's five-year investment five-year harvest. So, generally, they're usually only into a deal for five to seven years. They're not thinking ten-year investment term. They're thinking five years. So, you see a lot of investment capital. This kind of design to come in and out within a 48–60 month period. And in that time frame, you might not realize the upside of all those things you're doing. But if you're into ten plus years, now you're going to see that return jump significantly because of everything going on around it, and so you see in like Tampa, such as Tampa Heights or Armature Works and so you have a group that bought that land. I'm just taking the time to really plan it and build it out and they're seeing the value go up significantly every year as they build an office building and get Pfizer in there. They build new apartments. They leave it up and then they're kind of creating this whole neighborhood that when it's all said and done, the value of that place is 10–20 times more than it is now.

Q: So that long term point, how do you sell that to your equity partners?

R: It's a good question and we learned it. You're going through this last one in that you know it was the first time we've gone out to raise money for a 12 year investment minimum investment. But what we learned in going through that process is a couple of things. Number one, when you look at it from an IRR, that long term doesn't look as attractive as like a three- or four-year investment. So, four-year investment into a build stabilize the cell might be 25%, but the IRR on a 12-year investment might be 16%, but now you look at the multiple inequity. So that 16% over the 12-year period is like a 6X return, and when you look at it, it's like a 2X on a four-year investment will probably more than that. When you invest in a project and then you get repaid in four years, what do you have to do at that point? You have to now find another investment for that capital. So what's that process? That's looking at deals underwriting deals vetting? Am investing in him? Pay the taxes on the one you just sold, and then reinvesting it into the next project. And then do that again for four years. And then you gotta do it again. So in the same time frame that we've you have one project you now have been in and out of three deals, and when we did the math on it, after you've paid your taxes, you've had, let's just say, on average, three to six months downtime between deals, as you're trying to find an underwrite new deals, the returns are better on the long term. Just investing in one good project. Then in those three. But beyond all of that, I think a lot of people forget the big value. Investing in real estate is the tax benefit of real estate. The depreciation and the fact that you get this income with the benefit of all those net operating losses that you carried forward to offset the taxable treatment of it. So, when you take into consideration the fact that you've got the tax treatment of real estate that you hold long term, the value that you create over that time period and the fact that you don't have to find new investments to invest in and we were able to make that case to investors to get them comfortable, as long as that they had no need for liquidity within that time frame. And that's the number one thing. Do you need this money in that window so this is retirement money? You've got 20 years out. You don't need it, you don't care to have it back for the near term. It's a good investment if you need to have this money back and you got to do something with it other than just reinvest it. Then, it might not be for here, but if your objective is a dead end and you want to have this money invested for that ten-year period whether one dealer multiple deals. We made that case to show him like comparing one investment versus making multiple investments over that time frame. And then we didn't even factor into that the risk of going into a new deal every time. We've all known that seen that in the past, you make a ton of money on one deal just to lose it, investing in two new deals.

10 Real Estate and CRED

A Broker's Perspective

Lisa Reeves

Q: We're going to ask you some questions and want you to answer them from your real estate perspective and from also being a cred teacher and being a CRED alumni. The first question is who are you and what is your profession?

R: My profession is a small commercial real estate agent.

Q: Can you discuss the real estate aspects of a CRED developer? What should a CRED developer know about real estate and how it plays into their development? What makes a CRED development from a brokerage perspective different than a market rate?

R: Well, the first thing is to find out what type of project they're looking for. You need to know things like density, if they're going for low income tax credits, and the actual location is very specific. You really need to be there at the beginning of the project, what type of funding they're looking for and who is their client and who's is your developers customer at the end. Because without knowing that, not every piece of land will work. Because from there, you have to look at it from two different sides: from a funding side because of the very specific nature of affordable housing funding and then from the market side are people, their customers, even who are going to be at that location. I could find you land all day long that could qualify for tax credits but if people don't want to move there, how good is that property? So, you have to look at it from two different angles. Will the funders like it and then will customers at the end like it. And what happens is they usually don't bring a real estate agent in until they're looking for property and that's really the wrong time to pull somebody in. They should be there at the very beginning. They need to see the interview questions for when you did your client customer analysis. So, there's a lot of it that the real estate agent brings to the table and a lot of people don't realize diversity of skills that real estate agents have.

Q: It's a good answer and we all know that the CRED projects can be multi-use mixed use. They can have commercial and residential aspects to it. You being a commercial and the residential realtor. You've done both. What's

DOI: 10.1201/9781003109679-12

the difference between if you did have a multi-use CRED project or development residential, what do you need to know in commercial?

R: Well, commercial is going to be more about rental properties if you're looking on the residential side, you're going to be looking at home ownership too. Slightly different. So funding two different types of developers. Most of the funding I would have to say out there for projects at the beginning. Send me a rental property if you are looking at residential, then you're looking at home ownership; most of that funding is at the end when usually is directly tied to the borrower or the customer as opposed to the developer, so those are completely two different animals. And most developers can't do both. They do one or the other, they're very specific. It's not like some of the retail level developers can do multi-family townhomes and single family homes and maybe they'll add a small condo project and a big master plan. You can't really do that with affordable housing. You need to be very specialized because the funding is so specific.

Q: You're a special realtor because you have a lot more information and knowledge than most of the realtors that I've run across very diverse.

R: I like challenges. In fact, my specialty is blended properties and legal non-conforming property.

Q: So, let's talk about that for a second because most community real estate developments are going to be infill properties in urbanized areas. If I was looking at an urbanized area and I already had a bunch of different things going on, got a piece of property, it may have a history to it, I'm thinking of doing residential, some type of apartment with some kind of commercial, walk us through the process of what do we need to look at on that piece of property? It may have had uses on it that may concern us. What would you as a realtor be telling us about that piece of property?

R: If you're trying to do a mixed use property rental, if you're looking at doing some project, the first thing you need to do is find out does the local government want you to be doing this? Are they asking? Second thing is, unfortunately for projects like this, they are really technical and need to find an area that's already established. The problem then becomes it's not in my backyard.

Q: How can a realtor help with some of the not in my backyard issues? So, you mentioned early on that one of the biggest mistakes a developer makes is not bringing a realtor in from the start. First of all, why do you think that the developer does that? Why do you think that they feel like they don't need a realtor from the start?

R: Because in regards to real estate, most developers feel the real estate agent is there for sales. They're there to help them purchase or they are there to help them sell the property at the end. They don't realize that the real estate agent to sell past a lot of things that are going on. They need to know about the schools. Most of them have a decent idea about businesses and the community and what is going on. Is there a new road coming through that's going to make it horrific? Or people to commute in for the next five years that could be detrimental to the value of a property that can also make it difficult for people who rely on public transport to get to the property. Real estate agents are kind of like a mom. You think mom is there to just make sure the kids eat and sleep and get to their locations, but there's a lot of stuff that goes on behind. It's all of the unpaid for stuff that real estate agents do that helps them give you a better understanding of what is going on, what is the community like, who are the people coming in. I mean, that's probably one of the best things that I get from doing some residential or commercial as I know who's moving into the area. I know the age of the people, I don't have to go out to the search engines and stuff and get demographics. I know a lot of the demographics already because I just sold houses to 15 millennials. There are things that real estate agents bring to the table that don't really show up because it's live action as opposed to five-year old questionnaire information, they're on their boots. On the ground, they're the worker bees. They can tell you why there's no honey?

Q: So, in that vein, a local realtor versus an outside, maybe national realtor, I mean, you already hit the nail on the head that the local realtor has this affinity for the community. Let's first let's talk about some of the points you made like marketing. The realtor knows the market itself, they know what the area looks like, who's coming in, who's leaving, schools, etc. Would you say that a local realtor probably knows the history of the land?

R: Oh yeah, that's extremely important. A lot of people don't realize the social works projects from the 1930s made massive changes to the lay of the land. There are creeks that were built over. There were different roads that were moved. People don't realize what happened? The old maps are gone, city halls had fires. Little things like why does every home and thus block say they were built in 1924? Well, because there was a fire. And to rebuild the records they drove out and they said that look similar up. We're going to say they're all from 1924. Guess what? They're not all from 1924, but that's what the records show us. So, why is my backyard syncing? Well, that was one of the public works projects. That's probably a smaller creek was covered.

Q: On the residential side, what does a community realtor not do? I know there are certain rules in place, but what does a realtor not do?

S: We have to follow fair housing so we cannot ask people about their family status. We can't talk about their sex or religion and also their sexual orientation. We can't steer people and we cannot follow the pattern of redlining.

Q: I totally echo your sentiment about you how broker needs to be there through a long process. The developer almost has to have one on retainer in a way, speaking as a former commercial broker, it's important to have that. So, many of the things you talk about are kind of standard practice for good realtors. Let me rephrase. How a market rate broker is different than somebody who is going to be doing CRED? Maybe a little bit more specific on that. What is their difference? Because they're both, like you said, a good broker knows.

R: The difference is going to be the understanding of community development and even CCIM teaches it as a separate course, they have a specialized course on doing tax credit, multi-family properties. The average commercial real estate agent in a larger firm may not even know what is real estate sales contract looks like. They're even more specialized than say, residential is like residential. You might specialize to a community. But in your larger brokerages, for commercial, they specialize not just to community but to a specific type. So, maybe they only do industrial. Or maybe they only do multi-family or they only do leasing and they'll do retail and office. But it's just leasing. The community developer is going to have a little bit of knowledge, enough to get them in trouble in all of those types of properties. But they're gonna also be trained in the residential dynamics of community of what is going on a little bit more at the local level.

Q: The CRED commercial broker or those that deal in that kind of realm need to be much more of a generalist.

R: Yeah. You need to be a Jack of all trades or at least know a couple of people, but you do have to have at least the knowledge to know when you are outside of your realm. With community redevelopment and urban development, I need to know someone who might be able to explain this industrial peace. I might need people to explain to me about this historic hotel. It's not just knowing real estate, it's also knowing a lot of people who do other things outside of what you do because you're constantly pulling in. This is an industrial piece. I got pulled this industrial person in because they're going to have to tell me the history of this piece. And what's been going on with it. So, you're sort of a hub as a CRED real estate agent. You not only need to know the community and what's going on and the dynamics of residential on the customer side. You need to also have the people you can pull in with knowledge base to help your developer out because it falls outside of your real estate specialty.

Q: That's very interesting because when I was in brokerage, I did restaurants and land and if somebody wanted an apartment building, I didn't even touch it. I sent it to somebody and vice versa. But in your case, you would still continue doing it, you just would bringing that expertise to fill in that gap for you.

R: Yeah, because the specialty of what is going to be built is still gonna be out there. You can't send that to somebody else who specializes industrial because they can't do affordable housing. They won't know how to explain the management program on the other side of owning this, and the specialties of income requirements or that part of the demographics. So, you still have to state I do the project, but you're going to have to start pulling people in, who can work the piece until it is in development.

Q: So, in a sense, what I'm getting is you're an asset class, just like industrial, retail, etc. But your asset class that you're dealing with is affordable housing kind of stuff, right? In the sense of saying part of that asset class is these multiple entities that come together and then fruition of a project in the sense you're almost like a developer but that's a developer would do juggle these different people and you would bring in. Developer would bring in somebody to build the retail side and somebody to build that. So, it almost sounds like your asset classes that and you're almost a developer in a sense.

R: Yes and no. Basically, when it comes to affordable housing, property doesn't come owned affordable housing. It has to be rezoned, so it had come in as something else. And the problem is that something else is usually something that's hard to fix. And to do what I do, which is to understand whether or not people are going to live in your property and for people whether or not the community will support your property, I can't do a lot of the nuances before you start. I can do the market and the end product, but somewhere in the middle, If the property falls outside of just your standard regular land, then I have to know when to step aside because most of the time I'm not really getting paid a lot. So, real estate agents to do community development I think do it more because they want to give back to the community than for the payment. And so that's I think where we differentiate from your regular real estate agents we're going to pull someone in, who most likely will get paid more money than we will. A person who's going to control the land, but our benefit is we get to go from the beginning of this project to the end. We may not be paid a commission, we may be paid a consulting fee more so. Maybe we'll be able to add ourselves into the sale transaction, but it's really not as much a paid position as more of a I can add this to my resume as I did this for the community. I made this project move forward and I made sure that they got the community support they needed and they didn't make a big mistake.

Q: It sounds like there's a lot of moving parts in the kind of brokerage you do, as opposed to being industrial dude or whatever. Why do you stay in it? Why do people go into your area and not just go become the retail or the apartment person at CBRE or something? What's the driving force? Why did you stay in it and not go some other route where you probably could make more money?

R: I would probably say most community real estate agents who specialize in this at some point want to do their own project. They're probably looking to get enough knowledge and contacts to eventually become a developer.

And there are a few ways you can get into this. You can go through the developer, through the nonprofit or through real estate. My choice was to go through real estate because I ended up falling into community development after I started real estate. So, once I saw it, and I saw what it did, and then I saw the mistakes that were being made in it, I made a decision that at some point, before I die, I wanted to put a good project together that makes a big difference in our community and helps out in my specialty. If you look at the nonprofits I work with women and children, so my goal is to at some point either helping nonprofit build housing for women and children or to at some point do it myself. And that's I think the reason why a lot of real estate agents get into community development because they want to do a project themselves or they want to participate actively in the development of a project so that they've done something. It's more for what you've done than what you make.

Q: Do you think that's the similar path that people fall into that and decide this is in my area, instead of outwardly pinpointing like I am going to be a community real estate broker?

R: I think you have to fall into it unless you come from a lower-income demographic where you grew up in public housing or you grew up in a community that had public housing. I could tell you I had never seen public housing until I was in my 20s. I didn't know what it was. I didn't know it existed. There was no way for me to say, oh hey, I want to work and help people find places to live who can't even afford to find a place to live. That's something you have to learn about and fall into and that I think is the biggest mistake of community redevelopment is it's not out there, and I think that's also the reason why we have so many issues getting community support is because we don't teach people about this. Everyone should have a roof over their head if they want one, and they should have a clean place and a safe place to sleep at night. It's not talked about enough. Maybe you hear it on the news about the kid who lives in the car, who was able to graduate summa cum laude, but no one talks about why that should never happen.

Q: I would echo that. I teach in real estate program at major which is very market rate, very legacy kids for lack of a better term. And there is almost an hourly push back to this kind of thinking. Almost outwardly like saying we're not here to help, we're here to make money. And I totally get it, and eventually some of them do move into this realm. But it seems like you said they end up moving in by chance or they fall into it. And unfortunately, these are the people going to the development realm.

R: Yeah. And I will tell you like a CCIM course, most of the people who were looking at it, you could tell the participation was mostly commercial agents who did development looking for a way to get rid of that piece of property that they still had from 2008. It was really interesting when I took the class CCIM on tax credits because it was a mix of young kids and white young kids, recent college graduates who were just getting into commercial in their early 20s, middle-age people who looked like they had been selling too long and they were probably the ones who were investors who had property. Most were either curious young people or investors with property that they were like I gotta do something with this piece of property. Maybe I can build community, some apartments.

Q: Are there any other aspects to being a community real estate professional that weren't covered in zoning history, networking, legal, and community advocate? Is there anything else out there that you can think of?

R: Well, I think it depends on the project because if you do a home ownership like when we try to do the workforce housing which may actually be a big thing over the next few years, especially if prices don't fall again, which I don't think they will, there's a really large chunk of people who are priced out of homeownership, so I see a lot of workforce housing both on the rental as also the ownership side being able to market know if those properties can be marketed to the to the customer and or be sold. So, you need to know the front and the back side of it. We don't do a lot more in the middle the development of it tends to be a little bit more on the developer side as well as government and all of and construction, but we definitely are heavy in the front and then heavy on the end. Because we may introduce you to the rental agent who comes in, we may have our groups of people who are already educated in the management and rental of affordable units because of all of the income requirements and the annual income requirements as well as if we don't do leasing, we may be the person who brings in a leasing agent because we know that side of the service. So, that is the area where we do help out in bringing people in if the developer is not already established with their own people and connections, we can bring in those specialties as well at the end.

Q: Is there anything that we haven't asked you that you feel is very important too CRED as a real estate professional?

R: No, but I will emphasize that it's really hard for the real estate agent to get paid, which is why I think you don't see a lot of people who do this unless they're getting into development. And that is the reason why I do a lot of emphasis in my presentation about you need to pay your real estate agent. You need to figure out how to get them paid in the deal because if you don't pay them, they can't keep helping. I think, the reason why a lot of developers don't have real estate agents in the beginning is because a lot of what we bring to the table is unpaid work, attending a community support meeting is we don't get paid for that. We only get paid either if you decide to pull us in as a consultant or if you tie us to the land deal and that's why I think real estate agents aren't brought in the beginning is because they don't want to pay them for the commission for the land deal. And I don't know how this can be added into the book, but if you find a really good real estate agent who is going to the community meetings showing you how to do a community survey an a customer survey to get input so that you're doing the right project, you need to figure out a way to put them in the real estate deal.

Q: So, what I'm hearing is that maybe we could structure for this is somebody I'm developer, I hire you to help me. For instance, there's a nosey side, I'm really giddy about in Tampa, so maybe I will hire you, is one, to of course you're going to get your cut when you buy the dirt, get the dirt for many. Secondly, though, is to fold you in as a profit partner on the back end to say maybe that's 1–2–3% of profit whatever on the back end you as well. It is a good structure, the sense because then it sounds like you gotta put in the work anyways, but that also becomes a little bit more than character really help get this deal through.

R: Yeah, the thing is there's no line item for the real estate agent on any other proforma as we do and there really should be. Because then you can bring that real estate agent in at the beginning. It's kind of like doing your own plumbing. That's great, but at some point, you're gonna end up paying more money at the end when you buy the real plumber to come in and fix what you try to do cheaply. So, I find that the line item that most developers feel is unnecessary is a real estate agent and they don't if they get the right real estate agent and I will admit there's not a lot of people who specialize in it because there isn't any money in it. But if you start paying them, you'll start getting really good real estate agents, who specialize in this because they can feed their family and they can afford their convertible. But because they don't do this, you have to find those really weird people like me who just have a lot of extra time.

Q: I'll say this that up until last year, I had a real estate agent that I met through practice. I didn't know him prior to us doing and buying a piece of property as a CRA. And I would have him along on deals and make sure he got paid as a consultant to us just because of the immense amount of land knowledge that he had. He passed away last year, which I cried. And having him on the deal, there was always something he would bring to the table that I didn't think about, or the realtor or the lawyer didn't think about that made it worth our time. And his background was orange Groves. That's where he started was buying and selling orange Groves. But orange Groves become developments and he knows a lot of stuff into a little stuff about a lot of things and hit a good eye for the land. Another interviewee made a statement that the dirt tells the story. He said that from a perspective of developing something, it a history and it tells a story. Does this say anything to you when we say the dirt tells a story as a realtor?

R: Yes, but I also say the community tells you the story as well and you need someone who listens and can tell you the story of the community. The land can only tell you so much but sometimes all you can see about the land is what's on top and the community can tell you what's buried underneath.

Q: That's funny you mention that because he told us that the Housing Authority owns an old cemetery. And that they actually do interviews with community members, and the only way they find this stuff out is through talking to the community because no one else will tell them.

R: No, they won't and you sometimes have to go outside of the community to people who were part of that community decades ago, especially since a lot of public housing is rebuilt on top of itself, you almost have to find people who were there at the original public housing. Public housing can end up being built on top of previous because once it's that, it's very hard to reconfigure that property as anything else unless the area has gentrified.

Q: The last question I asked, you already answered and that was you've been a CRED alumni. You have taught real estate at CRED. You even come back to audit CRED. Why do you invest so much time in CRED?

R: It comes back to my core belief. I've always believed that there's enough for everyone and the more people who know, the more business you get. I mean a lot of people try to hold their information in and they don't give it out. You can only secure your position for so long. There's always someone bigger, better, faster, prettier that's going to come through. So why don't you teach them the right way to do things and give them the right information. Spread the knowledge. You never know what that person can do later on, so I've always been a great believer in teaching people, teach a man to fish kind of concept. I just think it's better for the community. It's better for everyone you see what happens when people horde. I think if you order your knowledge or just the same thing happens as if a person who hoards stuff. There's no pleasure in it. There's no benefit to it. And when you die, it gets thrown away, so why not?

Q: I'm glad you brought up the point of the pre-development. I teach it entitlements class and the very first lecture, the very first slide I give is a picture of a pig and I say you're all capitalist pigs. Gentrifying my neighborhood and that's how I start my class and I said it's your job to tell them in you're not and figure this out and then you go on to say that 80% of your profit are the deals made in the pre-development process. It's easy to go sticks and bricks. It's really not that difficult once. But all that pre-development getting with the community and I haven't been these community reports exactly like this and it's rezoning and all this kind of stuff that's important in understanding getting to know the community. So I'm glad you brought that up.

R: I'm very impressed with what I've been seeing lately. There's a lot more listening going on. And that's what I do as a real estate agent. I sit and listen.

11 The Role of the Housing Authority

Leroy Moore

Q: How has the business model of housing authorities changed over the decades from the original model to what you do now?

R: Certainly a great question and it has been incredible transformation and housing authorities both authority as well as activities over the years. Housing authorities were originally created to be the public housing owner and operator for low-income public housing and cost Section 8 vouchers. So, housing authorities as their primary bread and butter is public housing and Section 8. Probably 25 years or so ago with the advent of more complex real estate development tools like the Old Hope 6 program, housing authorities started entering into a world of mix finance and developer relations that over time, actually for some housing authorities who actually gravitated to that higher-level complexities became developers and became experts at assembling capital from various sources to finance and undertake more complex real estate projects. So, they became a little bit more resourceful than just the owners and operators of public or subsidized housing. Even market rate housing, and then shortly after that mixed use developments that would actually incorporate opportunities for jobs and just a more complete community development approach. So, the Tampa Housing Authority about 22–23 years ago made that transformation. We step forward and actually embrace mixed use, embrace mixed income, started really assembling a team and also hiring individuals that had expertise on the private side that could actually create that depth within the Tampa Housing Authority and we are only one of maybe a dozen or a couple dozen housing authorities in the US, has an extraordinarily diverse revenue stream as a result of that today. I often quote the number of about 20–25% of our revenue comes from the US Department of HUD for Section 8 and public housing. We've diversified our revenue stream extraordinarily as a result of new business ventures that we've gotten into being able to acquire housing that is not subsidized housing and provide that to the local community as affordable and workforce housing. From a financial standpoint and a financial dependency standpoint on that federal subsidy, and that's been key for us primarily because funding for public housing and funding for Section 8 to a lesser degree has always

DOI: 10.1201/9781003109679-13

been a struggle. It's always been part of domestic spending that unfortunately does not have both sides of the house full support consistently, and it's always been as one area that can be underfunded and historically public housing has been underfunded for 50 years or more. So, it was important for us to diversify revenue streams so that when the federal government cuts back we're able to not only sustain our services to our local community, but actually in increase. And having our revenue 75% independent of public housing and Section 8 funding has allowed us to do that to a huge degree by giving us revenue. Last five years we have not pursued a Housing Trust at all, because we've found ways of actually getting our very complex multi-phase projects developed without relying on a $35 million federal infusion. But by breaking it up into smaller bites by taking advantage of tax-exempt bond debt and a 9% tax credits and workforce housing opportunities and using our Section 8 program to cover more debt. So, a variety of ways and then beyond just building new affordable housing, we've been very competitive with creating up the revenue streams from renting office spaces renting warehouse spaces, assets that we would buy and actually rent to commercial owners. We've got restaurants and other commercial renters now that are generating revenue back to the Housing Authority. And probably the biggest single most important revenue stream for us is through our North Tampa Housing Development Corporation, the statewide operator for the project-based Section 8 program under contract we've consistently won the contract for about 16 years now. That we've maintained that contract for the state of Florida, which is about 43,000 units of project-based assistance, and we generate a revenue profit from that. We actually won the US Virgin Islands about eight or ten years ago, and we administer that program as well. We've actually competed for other states. We actually won the state of Georgia, but it's been held up for many years and now in protest. So, we compete for other states' activities. Primarily like I said so that we could actually generate a profit, use that capital to plug gaps and development pro forms to build or acquire new housing or other facilities that will generate independent streams.

Q: So, do you ever work with the City of Tampa with their CD CDBG funds? I mean, they probably are small potatoes for you guys.

R: Yeah, but no absolutely. I mean, city Tampa certainly is a consistent partner of ours and closing gaps and also given us the local match that we need in many instances for competing for state funding or federal funding. So, CDBG is a program that and I can remember 22 years ago up until maybe eight years ago, we routinely received a half $1,000,000 allocation from the city in the form of CDBG for our senior elderly housing program. In fact, not just ten years ago, I would say up until about four years ago when we converted our last public housing community, senior elderly public housing community, to RAD. For all those years, we were automatically going to get a half $1,000,000 and we would always put it into improving the conditions in our elderly rental public housing portfolios. Beyond just that, half-million dollar year commitment, CDBG has also been used to close gaps or give us our local match, take a $75 minimum match in some cases for a 9% tax credit deal all the way up to maybe a $600,000 gap that sometimes comes from the whole program, but sometimes the city can augment it with the CDBG program as well. So every deal, the city's had some money in it.

Q: Do you ever partner with the Florida based community redevelopment agencies in our increment revenue?

R: Certainly, the CRAs we have and I'm trying to make sure this is accurate. We've never received funding from either of the CRA. I believe that's accurate. We have always worked in conjunction with the CRAs one from a support standpoint to get the CRA is created to show that what we're going to be doing in the CRA is going to actually build incredible value because So, the increment and the ability to capture that property tax revenue for the 30-year term of the CRA. So in showing that and showing the lands that we're going to be developing within the CRA. Then giving the city and the CRA board I think a little bit more direction in terms of where they could put in public services like parks that benefit the whole community, not just our development but also benefits to our development. But we've never been a direct recipient of funding from the CRAs, but we've actually helped plan and organize and sustain because our properties when they redeveloped. And then, I consistently make this point how properties are paying property taxes and that's something that never existed previously. When there was 100% public housing, it was off the tax rolls. So, a great example always uses our Central Park Village community, which is now Encore Tampa. And that's literally 28 acres. That 483 units were 100% off the tax rolls because 100% of that 28 acres was all public housing rental, 483 units owned by the Housing Authority and never paid a single dollar in property taxes. But as we redevelop that 28 acres, it became 12 city blocks of development. And more than replacing the formulated 3 units with 667 total units now of affordable and workforce that sits on that same site. So, 200 units more than ever existed. But those are now mixed income and mixed use buildings, so the ground floor is retail restaurants and office space. Each one of those buildings, the market rate units in those buildings, our SAS for property taxes as well as the commercial space or assessor property values, so we're contributing to the increment even though we did not require any of that TIF financing or bonding in order to develop a public infrastructure that we needed for that site. The city, however, did develop parks and roads to lead to and other community public amenities that make that site work for us.

Q: I think this is pretty fascinating. The shift in housing authorities to this more kind of developer stance and what you're kind of doing is really what CRED talks about. Now you talked about the financing part, how do you handle your market rate? Do you go out and find private equity to help fund the market rate side of it. So, how do you handle the market rates side in connection with affordable housing?

R: We don't typically receive equity through REITs an even foundations. But how we end up with market rate units. So, if we wanted to do 20 or 30% of the units on a particular site with market rate, we pro forma them to be able to service the debt that we need for that. So, that's private debt we borrow in order to do that for commercial uses. It's usually part of our gap filler. So, we can pro Forma 100 unit building that has 70 affordable units then affordable units the pro forma for that shows tax credit equity and tax-exempt bond dead and public sources the market rate units then are servicing traditional debt. And then the commercial units typically is the gap, there's user gap. Also in affordable and some market rate units, but it just contributes to the gap the financing gap that we then have to close with the equity, our own non-federal equity, non-federal dollars, that we put into took deals. But a big source of private equity in all of our deals are low-income housing tax credits. And those are sold sometimes overeat I guess. Could purchase tax credit equity, but for us is always been insurance companies or pension plans or banks. And then within those organizations, they may have them sprint too, other investors, but it's used at Bank of America, RBC or one of those entities that are equity. But the market rate units for us are they there was we can build market rate housing. What we can't build without some public support and public sources is housing that we have to rent at rates less than our cost of capital essentially. So that's part of the magic, part of the deal is very different. Nothing is cooking caught cutter when you look at the capital stack. Our typical deals have a minimum of maybe 11 sources of funding. We've done deals as many as 15 or so sources of funding. If I was doing just the market rate deal, I've got two sources that I had to Jenna, so it's a lot of brain damage and each one rose sources of funding. If it's Federal Home Loan bank or if it's the CRA, the city, or all these other different capital sources, they all have their own underwriting standards and they all have their own attorneys. So, it's an incredible orchestra to conduct to get one of these deals closed, and then you don't repeat it, but you do it again with a whole different set of parameters. So every big deal really is unique.

Q: What kind of advice would you give somebody who might want to go into this? I mean, what should they learn? What should they do? What can they expect? Obviously a lot of late nights.

R: Well, certainly you gotta be a studious kind of person. But this is a lot. This is more practical education than dairy. It really is. It's working with different teams, spreadsheets, etc. My degree was in architect engineering technology and it wasn't the straight. Architecture program wasn't a straight engineering program. It was in the College of Technology and those guys that taught in the College of Technology were practitioners. They were the architects. They were the engineers. They were the contractors, the real people doing the actual work. And when you learn from the practitioners, I think it gives you that extra ability to really juggle in and do it as opposed to the theory taught subjects. So, that was a big part of both my formal education as well as my career development, being part of organizations like Urban Land Institute. Probably, if I stacked up all the organizations and I've been, in an alphabet of organizations as a member or have been a member of and involved in the ULI, I would be the most impactful career choice that I made 22–23 years ago. And because you're in the room with developers in, you're in the room with the finance people, and you're in the room with the attorneys in, the land use people, and the actual people that deal with the transactions, very transactional opportunities for people to get involved in. But being involved in those careers circles and just gleaning from everybody that you meet with. So, it's not a textbook career. It's a practical applied career. So, my best advice would be to join the organizations, participate in the organizations, and learn as much as you can before you start applying.

Q: Would you say, go to work for Housing Authority first, or go to work for HUD or go to work for the federal government or something along those or a city government and then come over? Or do you want to go work for a have them work for Trammell? If you were to hire somebody, let's say, would you rather look at somebody who worked for like a Trammel Crow or a Greystar as opposed to somebody who worked for like the Housing Department for Orlando.

R: If I'm hiring a developer, I want to hire someone who's actually seen it from the drawing board to on site walking in the muddy, dusty debris to seeing the people moved in. That have that real world experience. But if it's a lead as a developer, then it really needs someone who has actually experienced it. You could have been a bricklayer from bricklaying union. You then came into the office and learn how to actually cost jobs and submit bids for jobs. And then you learn the business side of it and learn how to actually run the business. To me, that practical experience is invaluable. So, HUD folks typically don't have the on-site experience. They've got the policy experience, and some of them are actual engineers, but there was from school into a policy position and are good people and that's what you need at those organizations. You need people who are good at policy. They don't necessarily translate over into a housing. And I know a lot of people from Hud, in Housing Authority operations. Some of them work great, and some of them didn't work out so great. But it has to be real transactional experience. I think to run a Housing Authority, Housing Authority gotta have that policy side but you also have to have that practical experience. You gotta know how to assemble things. You gotta get your fingers dirty and understand taxes even and how these indicate what that investor is going to be looking for, because in the 15th year when you ready to exit certain members, you gotta understand what it is they need to get out of the deal. You gotta be able to understand the other side of that balance sheet and what they are motivated by so that you can protect your interests.

Q: I notice that most of the projects you're doing are redevelopments of existing sites that the Housing Authority has had in the past. They've owned them for a very long time. You come in and you not only upgraded them, you have completely redeveloped them into a state of the art facility. How does the Housing Authority handle gentrification in the case like that? That's a very important question. I know you have a great answer for it.

R: Yeah, it's just such a buzzword these days and it automatically sends people to the worse applications of what gentrification is. Gentrification happens all around, and it really is a real estate term and you've heard some people use that. Don't be fearful of it. Learn how to manage the impacts. It's not avoidable if you're creating transformation of projects. So, if at the end of the day, you want it to be a transformation project and what that means for us is a community that is diverse, that is inclusive, that is prosperous, that has accessibility to jobs and recreation, not just housing. If you want that type of community, gentrification is inevitable. But you can manage the negative aspects of it. We can influence, but we can't directly control what goes on around. And that's the biggest part that. I tried to communicate outside in the communities most of our development because they are redevelopments. These are sites that are housing existing people. For tremendous amount of terms, years and decades and communities that have seen disinvestment or tremendous amount of decades and surrounding neighborhoods typically have suffered because of that lack of public investment, private investment. So, that's opportunity. It's not a problem. That's opportunity. Before we move the first family we're in that community in some cases for two and three or more years talking about what we are planning to do. Not just to our community, so I'm talking about in the buffer community outside, on our land, are the way we come back? So from a social economic standpoint, gentrification is not the displacement of one racial or economic group of people for a different group. That's not the result of our development for our site because the people that come back or either the people that were there or people who look and earn and the profile is the same because it's one way to do this. But where that does happen viciously is in a surrounding community where housing, that it could be rental housing or it could be owner-occupied housing, they see the activity and the prosperity taking place and understanding that is going to grow in that community. And they're predators that will prey on those communities to purchase those properties and then redevelop something that the people who once lived there can't afford any longer bottom line. So, there's a need for us to make that eyes of awareness in the community. And when I'm in the community, I'm talking to business owners and I'm letting them know that we're going to be moving 1,500 people that today come to buy soda from you or loaf of bread from you. It's gonna happen. But from a commercial standpoint, and there's commercial gentrification as well as residential, some safety small business owners understand that why this community is being developed. There is tremendous opportunity for you to sustain your business. The families that were moving out, or indeed leaving. But we're going to have hundreds and thousands of daily workers on that site every day. And these are people that have disposable income earning good salaries. So that's an opportunity for you to perhaps depending on your business to tweak your business model for this period of time. There's an opportunity for you to market too. A temporary workforce and then letting us know what the permanent population is going to look like coming back in. We have 400 units we are going to have 600 units of low-income individuals, but we're also gonna have another 200 or 300 units of market rate individuals. So, what you're selling today might not be appealing to this population five years from now. So, there is a need for you to do your own business planning to understand how you can sustain doing this transition and then how you can supply this community. But I've said to people specifically in meetings if you want to stay there, we want you to stay there. This is what we're going to build on that 2/3 that we do all. But if you want to sell, please come to us first because it allows us now to complete that block. But if you don't live on that third of a block that we are interested in and you live across the street, or you live in the broader neighborhood and you're on your property, don't sell your property. If you want to participate in the upward prosperity of this community, and if you get a daughter, get somebody who can actually sustain your family ownership of that process, a matter of just letting people know being, for still there are predators that come in and People lived in a home that maybe they paid $20,000 for 30 years ago 40 years ago somebody comes in and offer him $50,000 for it and they wanna grab that $50,000 and then they realize they can't buy anything for $50, and then at $50,000 house now becomes two homes in the lot value alone is $50,000 for each of the two homes. So, I just sort of explain the nature of real estate and what the opportunities are and also the fact that you're gonna be hearing from a lot of people trying to buy your property. And if you want to keep it in your family, find somebody in the family. If you really want to sell, come to the Housing Authority we might have an interest in it because what we're acquiring is usually for purposes of providing affordable workforce housing. So, at least we can try to keep the mix of working-class individuals in that community longer.

12 Mosiac Development in Action

Roxanne Amaroso

Will you please state your title at your work, who you work for and just give a brief description of what you do?

I am one of three owners of a company that I built, called Mosaic Development. We focus on multifamily development. Three to five story and master development in public-private partnerships. What that really means is that we work with cities who want to redevelop land that they control and have certain products and we help them envision what works for them, and as long as it is inclusive of apartments, we are their master developer.

So, the first question I have for you is what is the difference between community development like with CRED and just normal everyday private sector development?

Well, I do both. Private development is where developer looks for land opportunities or entitled with the appropriate zoning for the use that they want to do. Some are entitled during a contract period and they're building it for themselves. Myself and my company with my partners are merchant developers, meaning that we build it to sell it rather than to hold it. In the public-private community development is where you would seek an area which has experienced disinvestment. Normally, it's in a very urban corridor of a city and you would revitalize that area by creating a product that the community needs. It addresses a core need and that core need then attracts other uses around it. And eventually, there are sustainable long term uses and creates a tax base for the city and it creates a great opportunity for residential uses and other applications.

Can you walk through just some of the bigger points, the 40,000 feet points of the process that you follow to do community development?

It always starts with research. You have to understand the area in which you are interested in intimately. The first thing we look at is the needs of the community. An example would be that there is not an easily assessable grocery store. And there is no affordable housing. And these are just simple

DOI: 10.1201/9781003109679-14

examples of what the needs might be for that community. And you then look at the vacant land and what belongs on it. Sometimes that's compatible with the existing zoning that's in place frequently it's not compatible, so you're taking on a re-entitlement process. And you model that on a pro-forma, which contains the cost to build the product. The interest on the debt and the equity, the impact fees, all the fees, all of the soft and hard costs, to produce an end product. It will strike certain numbers that will attract investors. So, there are hurdles. One would be an internal rate of return which simply is the interest rate that an investor gets on their money at a project level. Another would be the profitability of the project for the period in which you hold it, which could be in my case be a two- to three-year event in other people's minds. It would be a five-year proforma and if everything comes together, you proceed forward with concept plans that you bring to debt and equity providers to see if you can assemble the financing to start building.

So, the big question is can you do community development without government assistance?

No, I really don't believe you can. I think that when you go into an underserved area or an area that's experienced disinvestment, there's usually a gap in the financial proforma. In other words, it's underwater. It's not profitable. It doesn't make sense. Cities have various funds, so do counties and the state and the federal government for that matter, in which you can apply for, that helps fill those gaps. Cities and counties are normally driven to want to invest for two reasons. Either it's an area of severe disinvestment which would equate to an area with high crime rates, high poverty rates, low employment rates, and those would be motivating factors for a city or county to want to assist with the financing. Frequently, there are competitive funds at the state level in which four profit entities would apply for and provided they do everything correctly within the application process, it eventually becomes a lottery because there are always more projects than there is money. My claim to fame is having 13 funding sources for one project. It involved 13 separate entities willing to invest their money, some of which they're never getting it back, and they know it, and you know it, but what they are getting or things like a tax based generation, employment opportunities, training opportunities for people who need to be trained for employment, it's a revitalization effort that takes an entire community to embrace.

One thing you've always talked about is risk and reward. Give us an idea of the level of risk that you take in order to bring a project online and complete it.

I'm gonna start by telling you that our team is considered smaller. There are four major individuals within the company and we have other individuals which are supportive companies; it's a sister company which employs about 100 people and we get our financial accounting done through that, our marketing, and our demographics. But in terms of the risk reward, we do three projects at a time, meaning three projects under construction while we are sourcing three to five more, and we look at about 40 deals a month and we pick one or two that were willing to investigate because they might have something. And at the end of the day, the project size for the cost of the project ranges between 40 and 50 could be as high as $60 million per project. So within that, think about that number for a minute. If you think about that number I have to personally sign a guarantee to the bank for the full amount of the proforma. And that's not just the bank, it's the investors too. I have to guarantee them that I am not only returning their money, but in the case of the investors that they get a certain percentage, which we negotiate with interest and it compounds because they're not paid until construction transfers to leasing and income occurs. I also have to negotiate with multiple banks and I use an outside consultant to do that to get the best rates, but in every case, every dollar which I take in myself and two partners signed the dotted line stating, we will repay it no matter what happens. So, I'm taking personally with partners 40 to $50 million in risk times 3 right now. But in this case, my book of business today is about $209 million. That's about $209 million in personal guarantees. I do not have $209 million. I can't afford to make a mistake. I have the blessing of working with great people, highly experienced, reasonable people and I use our best efforts to mitigate risk to the best of our abilities. So, at that point, you have to ask yourself, am I willing to take that risk? A bank has the ability to try to make itself whole in the event that there was a problem which resulted in a default. And the bank has the right to take everything you own, but the house in which you live in and is homesteaded. Everything. So, that's a lot of risk. To mitigate that, we make sure in our loan documents that we have a period of time where we can fix the problem. Whatever the problem is. In our equity documents, we are guaranteeing a return on the capital with interest. We are not guaranteeing the profitability because it varies according to market changes. We try to be very conservative in our underwriting so that an investor is thrilled with the outcome rather than angry because they didn't make the money you promised. I would say to anyone and I hope you don't mind me adding to it. You want to make a difference and want it badly. You have to take an educated risk on anything you do anything in your life that's truly worth having you have to work for. This is no different than that, but you're putting your wealth on the line to do that. And everyone else is along with you. And when everything is said and done, you have one reputation. In other words, if you screw it up once, besides the obvious horrible effect it has on your personal finances, you have a business red flag on you for being someone that someone wants to give you money to invest in a project.

OK, we've talked about the risk that you take now. I want you to talk to me about the reward. What is it that makes you go out there and do this?

I tell you the first answer is pride. I take great personal pride and building something meaningful for a community. And knowing that I have the opportunity to do legacy work. Frequently, I take my children, my spouse, and my grandchildren to terrible communities and say this is what it looks like now and they come to the grand openings to see the final product. So, pride is worth a lot to me. It matters to me, impact really matters to me. Financially, what a blessing if you do a good job. I went through a period where I decided to take a risk on myself and we built this company eight years ago and it's been life changing for me and for my family too. But it absolutely embodied risk.

Would you please give us the story of how, all the way back to high school and what you did? Because this shows that you don't have to be born with a silver spoon in your mouth. You can make your own way. So, please tell us your story.

I'm 62 years old. I am the youngest of seven children in a family. It was a big Italian Catholic era. All families were big families. My parents divorced when I was very young. And my mother worked several jobs at the same time to support her children. I didn't get to graduate high school. I got a GED and I went into the Air Force at 17 and I had to get emancipated to do that because I wasn't old enough to go into the service otherwise. I wasn't ready emotionally for college and I didn't have a dime. Going into the service gave me a great foundation for a work ethic that would allow me to succeed. I'm personally a very driven person. Once I emotionally commit to something, and I'm very careful in what I commit to, I take it all the way and nothing will stop me and it takes that kind of mentality because some of the problems that occur in a course of development are very hard to overcome. They take creativity. They take guts, simply you have to be able to look someone in the eye and promise them certain things and then you have to deliver on that. I will tell you that over a period of time I learned that I absolutely loved and had a huge passion for the multifamily arena and the master development. I guess I'm fairly intelligent background, but very poor educational background. So, everything I learned was really out on the road learning as I took on different jobs. I'd always had a passion for real estate and I started out in my first development company at $8.00 an hour as a secretary. Today, I have the privilege of saying I think this is my 15th year with CRED, so I am teaching at a USF at accredited course. I manage a book of over $200 million. It has changed my life and I didn't start that until I was 56 years old. And I've just gotten divorced. I took on a whole bunch of debt. And I bet on myself. I wanted this so much. And I was willing to fight for it. A long time ago, my mom told me something that was very meaningful to me. She said you don't have to be the smartest. You don't have to be the fastest, you just have to want it more than anybody else does. And I'm that person. And it served me well. Today, I went from a serious debt position in my mid-50s to my house is 3 quarter of $1,000,000 and it's paid. I've earned significant rewards when a property sells. And my base salary is about 125,000 a year and I'm blessed. I take nothing for granted. And everyone's voice matters from the poorest person, the least educated. Everyone has something to contribute. But if you have a passion and a desire at a drive to take it or you wanna go nothing can stop you.

So, when you are and picking a project, you begin to put together your proforma. And first of all, how long has your proforma ever stayed the same? Or has it evolved with each project with each economy that does it, even today after all these years of doing it, do you still adapt your performance?

Absolutely it changes. Here's what I would say to that when I start a proforma, if it has the right metrics in profitability and sustainability, if it's a money maker and it would be attractive to an investor and attractive to a bank and I'm confident it will work, then it becomes a matter of refining that proforma and the way to refine it is validating all the numbers within the Proforma I typically work with a general contractor and an architect. The contractor will get hard bids from his subs, his top subs and they'll give me an estimate of cost. I signed two kinds of contracts I used to sign an AIA, which is the American Institute of Architects contract form, which was a cost of the work plus a fee with a guaranteed maximum price. I always thought that was a safety net. It's not. We've since moved to a stipulated one, so it's a fixed number where the architect and then I picked the architect and I say myself, it's really myself and my partner's. We don't make decisions independent of the other. But we picked the people that we know. We understand their history and their capabilities and we trust and they actually signed contracts to go under the general contractor who's going to make absolutely certain that he hits the numbers and one of the things that's really important. So, there's not a big spread in the numbers. Later is making sure you give him what your expectations are. Is he going to give you a $500 wipe small refrigerator? Are you gonna buy stainless steel and full sized? How tall are the ceilings? Are they 9 foot ceilings because they should be whereas a tax credit deal might have 8 foot ceilings because they couldn't afford to build it higher. There are variables in cost. You estimate cost of legal, environmental, the actual construction you're hedging your bets against lumber cost increases which we got hit very hard this year. I had one project alone that was $2 million over budget Just in lumbar and right now lumber and sheetrock and availability of getting materials is very difficult and costly too, and the longer the contractor sits on that site waiting for those materials, you know it's easily 100,000, 200,000 a month, for every month that they sit there not being productive.

In other words, it never costs which you think it does. It always cost more so when you are truing up these numbers, we've gotten to where we grow our contingency money and a contingency is simply for things that didn't go right. And it could fall into the hard cost bucket, which is built bricks and sticks if you will. Or it could fall into the soft costs budget, which would be architectural design, a civil engineering, any consultants that work for you generally and legal and legal swings vary widely. Attorneys bill in 15 minute increments and a really good attorney charges 4 to $500.00 an hour. So, we cost estimate and then we validate all those numbers and we force our legal counsel to give us a budget for their selves because they're more apt to adhere to it.

As we validate these numbers, we do update the proforma and an average deal for me could have easily 17 versions of that based proforma. It doesn't materially change. You're always within a certain zone, that being. If you write a conservative Proforma, you don't overshoot on the rims. You don't underestimate the hard costs and you stick to the plan and you don't allow anyone in your group to make crazy decisions about how lavish something should be or just making poor decisions. The profitability is somewhere in the 20% range which are huge numbers and after the investors and the bank it paid off, there's actually a very nice payroll for the people who worked so hard to make it happen.

The principles we get paid last. We get paid completely last even in the bonus thing that we do for our internal folks who are not part of the partnership. And it's important to them. They have families to feed too. And they're taking risk to, and if they're engaged in a manner where they are bonused, they put some real effort into making sure you're a success, so for me the trick to mitigating risk is adhering to the proforma, updating a proforma and this is before you go to a bank and lock in, validating every single number in there to the best of your ability.

And then you're simply betting on yourself. You're betting on interest rates. You're betting on construction timelines, you're expecting certain things to occur at a certain time, and when they don't, you have to shift. And everyone shifts with you, so the proforma is a vital tool. It is the heart of development. And it's something to take very seriously, because when everything is said and done, you are going to sign your name, personally guaranteeing that money. And if you and your group of developers in your team do not have 15 to 20% of liquid cash, your bank accounts of the loan value, You're not getting a loan. So, if I do four or five deals at a time I sometimes have to go buy a guarantee from somebody else. And the cost of a guarantee typically runs me anywhere from 200,000 to $1,000,000. And it's underwritten in the proforma. A bank will allow you to take the totality of a proforma, the totality of the entire cost of the project, including interest and lease up costs and things you don't think of when you're building necessarily, and you're limited to that amount during construction. So, our equity raise really is from 10 to I have one deal right now it's 22 million and just in equity. It's a $60 million deal that will sell for $90 million. When it's completed in about 29 months. Big numbers.

Will you talk a little bit about properties that you picked that you knew going in were brownfield sites and properties that you went into and at the time that you didn't there was a brownfield issue, but discovered in the process.

Sure, I actually love doing brownfields and so did my partners. Over the years, we've become very adept at partnering with incredibly intelligent and resourceful consultants in the environmental arena. During due diligence on the land, which is during the contract period where your deposit is soft, it's refundable, we go out in the field, and we pay for the testing to make sure we know what we have. And there's always some kind of awe or oops somewhere. It's the what if factor. The environmental land is especially appetizing to us because you have the ability to take something that's not on the tax rolls because it's not usable and its present condition, and it's posing a health and environmental risk to the entire neighborhood that surrounds it. Cleaning it up, and there are funding sources out there that will offset up to 60–65% of your cost. So, we have the ability to recoup the money and I frequently take the environmental work and I put a bridge loan on it, which is a shorter-term loan. It still gets the same interest rate as my equity does, but it's about shorter-term loan and as soon as the Department of Environmental Protection Agrees that we completed the work in the manner that we signed a contract with them, it's called a ground field site rehabilitation agreement or re remediation agreement, and as long as we followed the protocol that was outlined and approved by the DEP, when everything is said and done and they're satisfied, we end up getting what's called a no further action letter. What that does is absolve us and any future users for any bad action that someone else did to contaminate the land. And I'm willing to take that risk because we collectively know what we're doing. And we know how to implement a plan that will make it either capped so it's compartmentalized in the soil, that is bad soil, under concrete or pavement, and we remove a portion of that soil offsite, it could be all kinds of different kinds of environmental work, but it's a funding source and a financial opportunity to take something that's negative to the city, negative to the residents, negative to the commercial end users and turn it into a positive and a brand new product.

How many years have you been with CRED now?

15 years. I served the state of Florida's Brownfield Association board for a few years, which helped me get a really finite understanding of what money you can apply for, what the rules are, and how to time everything. Because there's kind of a waltz that occurs during the development process. You're dancing with timelines. And there are multiple timelines and deadlines and provided you follow the path and it's your path because you help facilitate it, you made that and your proforma dictates that. So, it's a healthy project. If you do that and you're in the right place and you're doing the right things, it's incredibly exciting. It is fulfilling emotionally and financially. Wow, life changing.

So, you probably have the most time in with the CRED classes, is there one message that when everything else is set, all of the financial, all the technical, all the environmental issues have been talked about, that's like this is what you want to tell? Is there a message that you like to deliver to would be developers.

There are actually a couple of messages. The first is about character. I'll go back to my mother again. Who said when you lie down with dogs, you get fleas. Be careful who you partner with. Make sure you know who you're dealing with. It's very much like a marriage in equity, particularly. They can make your life a living hell or they can be an awesome partner and it has everything to do with their reputation and their capabilities. So, whether it's an architect or an engineer, be mindful of who they are, what they've done, and what their reputation is and when they show you who they are, believe them. That's the first thing because you are mitigating risk by doing that. The second thing I would tell you is development is not for just anyone. It takes a strong-willed person with an emotional commitment to develop exactly what they promised because you're making a promise to the city to the neighbors. You're making multiple promises that matter to people. They're life-changing events. You can't afford to make a mistake. We all make mistakes. I like to make new ones. One of the things I would tell you is if you want to be that honorable person, that person that's willing to be daring enough to make a change that affects other people in a meaningful way, a positive way, then you have the guts to go forward. If you can't look a city in the eye and say I need your help, this is what I want to achieve. My proforma says I need $1,000,000. You will be the 5th entity in line to get repaid. I don't know if that's gonna happen or not, but I need you and here's why. If you don't have the stamina to do that, this is not the business for you. So, conviction. You have to have conviction in what you want to do. I think that's probably a lot of messaging, but they're the most important ones.

I think you're right. I think it comes down to do a character because development is not easy and it's not for the faint. Is there anything else that you would like to add in this interview?

I would. I think the general public and your world is a developer going back to engineers and architects and contractors and all of the people that you partner with. They make you who you are as a developer. They greatly contribute. They also have the ability to fall down and be weak. People fundamentally want to do good. They want to do the right thing. They actually want to support you. I used to laughingly say in the tax credit and bond world, no does not mean no. No means baby, you have to have the fortitude to say, I understand what you're telling me, would you consider doing it this way? Hearing what the need is of that money, a city or county It's job creation. It's sustainable housing. Understand what their needs are and make sure you address them. It's how you create those opportunities and they really are your partners, all of them.

13 Using Incentives for Affordable Housing

Reed Jones and Ralph Settle

If you could just first introduce yourselves.

Sure, yeah. Ralph Settle, graduate of the Clemson MRED Program 2009 and presently I am a principal in Equity Plus and a principal in Beacon Property Services here, located in Spartanburg, South Carolina, with a focus on acquisition and development of affordable and workforce, low-income housing and property management as well.

Reed Jones, I am also a graduate of the Clemson Master Real Estate Development Program. I am a development project manager for the equity plus family of companies where I focus on acquisition, development of affordable housing, workforce housing, among of housing deals and construction of those projects.

So, how do you see the kind of affordable housing, community-driven development to be different than the market rate real estate development?

Well, just, you know, first and foremost, for us, it takes just a higher level of coordination between all of the public partners and your obviously private equity partners and they need it with deals we're doing here locally and you know, one that that one we're pretty excited about as Robert Smalls here in Spartanburg is the replacement of 190 units that have been kind of the worst of the worst, hardest to house conditions in Spartanburg for a long, long time.

And to put it into perspective, when the then new assistant city manager, who is now the city manager in 2007, the number one issue between the city manager and the assistant city manager. What they wanted to tackle from a community redevelopment standpoint was Norris Ridge (present site of Robert Smalls). Now is 14/15 years ago. It was built in the late forties, early fifties, so post World War Two.

But that really would not have been able to be accomplished if it had not been from a high level of coordination between City Council people, the city staff, County Council, the county staff level, the Economic Development

DOI: 10.1201/9781003109679-15

Agency here, which is underneath one Spartanburg. So, where it is housed underneath the chamber. It just took so many moving parts mean moving people to be able to get where we are today. Ultimately, we were able to get a site that had been sort of pigeonholed from the city's perspective as a pressure relief valve of something of some sort of magnitude.

And so when we purchased Northridge in 2020. The light bulb went off because I'm local to Spartanburg. I've been in Spartanburg my entire life other than the six years I lived in downtown Greenville. I've been very active in the community, among a lot of boards and a lot of people in city know me from prior to that. Northridge was an out-of-state landlord and very little community involvement, if any. Nobody really knew who this person was or what they did other than Norris Ridge.

Well, fast forward, we purchased the Northridge asset and we were able to get 23 acres for $100 so that we were with a lot of strings attached. Obviously, as many, you know, public-private partnerships have with their development agreements.

Along with that came opening up the specials for special source revenue credits as it relates to the sources from a taxing perspective that had to be passed not only by the county, but by the city and vice versa. And so there were a lot of moving parts to make the deal work. Ultimately, we were able to make it make it happen. But in a market right deal, more of my best friends who graduated the embed program in 2007. They do market rate deals all across the Southeast, the Midwest, mid-Atlantic and they it's just purely a different animal.

They go in, they find the dirt in an entitled Build it and sell it, you know, 30 months, which is a great model. It's just not these models. It's just a completely different bear.

So, I mean, with that regard, what do you see? What separates kind of. He set it up really nicely for us. What separates a kind of developer you guys are with, you know, the market rate developer you just kind of talked about and what it's kind of that. Dana? Yeah.

Well, and so I think from a philosophical theory sort of way. No, no, right, wrong what you know, right or wrong way to do it. So, I graduated with a program that went straight into building high-end luxury hotels in major markets, you know, Chicago, Santa Monica, Napa Valley, you know, Bryant Park, Miami. So very, very fine. Very, very interesting. Very, very intricate development deals with NATO development here in Spartanburg and in the background, I was always on community boards. I was on the habitat board. I was on the Housing Authority Board. I have a passion for the folks that need affordable housing that need that extra leg up. And I was doing a day job and being on boards at night. And ultimately, what I think sets us a little bit apart is that, you know, you have and read heard me say this, you know you, you have sort of passion meat expertise, right? So you know, we know the development process inside and out.

But now we're able to effectuate positive change on a population that needs someone to do this type of development. And you can always use this as well as you can. You can do good for others and still do well for yourself. So that's where we're at from a day to day perspective where we think, you know, it's while we have Beacon Property services is because not only do you build a development, you go through the development process, you get the CEO and then you get the folks occupied, either through voucher transfers or through a, you know, a lease up to a workforce deal.

But then we think the next piece of the puzzle with Beacon Property Services is the property management side is that's where you really continue to serve the population. So, you know, we like the idea that, you know, we know who's in Unit 221B and we know that, you know, they're handicap or they need to be moved to a handicap or, our tenant base as best we can. I mean, you know, obviously the property managers know them better than the executive staff.

But we do have that hands-on approach where we take it to the next level of serving the population that we build for.

One thing I keep hearing coming out of what you're saying is this passion for a kind of service to the community? So, what how would you describe you guys being successful? I mean, it's pretty easy on the market, right? How do you rate success from your end?

Sure. And, you know, I'd love to get Reed's perspective on this as well, but you know, as much as I'd love to be a philanthropist and not have to worry where my next paycheck comes from, that's just not the reality of the world. So, these have to make economic sense not only from just the development, but there also has to be a little meat left on the bone for some people to eat as well. So, we don't do it. You know, for charity, these things do make some amount of money.

They don't make as much. But you know, it's you have a lot of things that make up the whole, I mean, you know, in one right, you know, we do make a property management fee and it keeps the lights on. I don't think anybody's going to retire to the Caribbean on that, but it keeps the lights on bright and it's and that's able to afford to pay for three, four or five different people that, at the end of the day, they get to take a paycheck home to their family.

So, there is that aspect of it when we're looking at deals, these deals have to underwrite. There are definitely underwritten differently, debt service coverage ratios are different. So, we have to, you know, these things do have to stand on their own from a dollars and cents perspective. You know, there is a development fee. It can be deferred and it often is, you know, there's enough to set that the front end to keep the wheels in motion. But you know, there are some deferred development fees, so that does help.

And then these things do cash flow. And so, you know, if you have your partnership structured correct and your capital stacks, correctly that do spin off some cash flow. So, it's not total philanthropy, but it's definitely not your 2x in 30 months, either.

Yeah. And I think another point to touch on is the fact that especially in affordable housing in the markets that we work in, it's very important for us too. Create a strong relationship with the communities that we're working in, it's not something where we're building this. We're not merchant builders. We hold what we develop and we stay in the communities and want to ensure that we're putting in systems in place for the residents of our communities to be able to find success on their own.

So, we really look at resident services across all of our portfolios. Another thing to touch on is, you know, the risk metrics are different for affordable housing than they are for market rate. So we take into account different levels of risk and different types of risk than you would if you were doing something like Roth used to do, developing luxury hotels or developing even market rate apartments. We have to look at different aspects of risk, whether it be community-centric, risk, community wanting, affordable housing, or not understanding what affordable housing really means. A way we've gotten past that, especially in Spartanburg, is ensuring that we're building a very high-quality product that will stand the test of time. So, in Spartanburg, particularly, we're building Robert Smalls using a modular construction concept, which not only allows us to produce a higher quality product quicker, but it reduces the cost of project because we are delivering it faster.

And so it cuts back on the back end interest that we pay on a construction loan. We're able to more closely track the development of individual units because they're being built in a factory rather than being set frame on the side. There are issues that come with that, but it's a tool that we've found really helps us ensure a quality product.

To follow up on some of those, Ralph, you kind of put it out there that you still need to leave some meat on the bone. What kind of incentives etc. do you tackle and employ in your development so that you you're able to do that and have some meat on that bone?

Sure. So, a couple of different things. I think one of the biggest issues that you run into when you're talking about doing an affordable deal versus a market deal, is something as simple as taxes, these things don't cost any less to build in a market rate. So, you might not have the same type of corner, you know, cabinets or the same type of countertops. But other than that, too often we're still a two by four. And so one thing that we did here locally with Robert Smalls prior to some legislation change is we implemented what I mentioned with the economic development entity underneath one. Spartanburg S.C. where we had a special Senate source revenue credits that went into the deal that we're able to put a floor. No ceilings on real estate taxes. So, based on a metrics of what this deal provides, I think about it from a true economic development standpoint. We live in the state of South Carolina in Spartanburg, we have the largest BMW plant in the world with Plant Spartanburg. And just like any other tax incentives to get some big, big, big, big, big, you know, manufacturing employer special source revenue credits are in place for infrastructure.

They're in place for a whole litany of different things and lots of times its taxes. They don't pay taxes, right? So, we implemented that same concept at Robert Smalls. They waited, obviously, you know, not jobs created an investment in the community differently, but we were able to implement a special source revenue credit that they took into consideration all the positives of this project. All the boxes that it checks. And they said, Yes, we believe in this, and I'm comfortable going in front of County Council and City Council saying because of all of this, one of the levers that we can pull is we're going to reduce their taxes to X.

So, those are not coming out of the optics line, right? So, we're still paying a small amount of taxes, but not near what we would be paying as it related to at market rate. If you bring, a non-profit partner to the table, your taxes are zero. And so with that being said, we have a strategic partner and non-profit partner in justice housing that does much of what we just said about residents and community services. They are a partner of ours in deals going forward in the city of South Carolina and due to we have zero taxes in the development deals based on that, having a nonprofit partner.

So, that's just one little way that you can eliminate some of the dollars that a marker a deal would be paying out the door that helps us meet debt service coverage ratios and bring some cash flow back to the partnership group.

Great. So, you know, here's a bigger question I'd like to hear from Reed on this one to a little bit. What makes a successful developer in your arena? And I'd like to kind of draw on Reed as you recently graduated from the MRED program, what kind of advice would you give somebody going into it?

It's a great question, I think, for affordable housing, a commitment to community is really where it needs to start. Much like what Ralph is talking about with Northridge Ridge, the property was built in the 50s. It was cinder block framed. It was your typical affordable housing. The first thing that comes to mind when you think of affordable housing and it was owned by someone who was not local, that was at the point where cash flowed for them.

They had stopped doing renovations, major repairs. It just was sustained without the commitment to community. The product was declining. So for us, I think it starts with a commitment to community and a commitment to helping and improving communities. I think anybody can be a developer. I think anybody can run numbers. Anybody can learn how to develop products. But without that commitment to creating a quality, lasting development that really is impactful in a community, it's just another development.

I think we're a very strong player in the affordable housing field. I think we know we build quality work and we put our hearts and souls into what we build. So, I think it first starts with that commitment to community. I think the second thing is flexibility, especially over the past 18 months, we've really learned how to ride the wave as things change day by day. But even sometimes hour by hour. The third thing is attention to detail, especially when you're using public funds for affordable housing developments. You have to follow a qualified action plan and you really have to be willing to get into the mind of a qualified action plan to ensure that you're meeting these requirements that have been laid out and set by the state. I think something that goes along with that is a strong relationship with the State Housing Authority in the local housing authorities because there are some you can't do everything, qualified action plan as laid out to create the best development, but that best development doesn't work everywhere.

And a strong relationship with those agencies allows you to have conversations that can make or break your development that can allow reduced unit sizes based on the floor plates that you're able to build that can allow waivers for something that you don't even lifts in units the number of lives required in units to create ADA accessibility. It's a strong relationship with those housing authorities and state agencies so that you're able to create a development that works for everyone because not every development can follow the same exact guidelines to a tee.

I think after you hit those three goals, it truly is a quote-unquote typical real estate development where you are doing the same things you do with any type of development. But I think that sets community housing and affordable housing apart from other types of development is it's a strong engagement to communities not saying that other types of real estate development, asset classes are not engaged in the community. I'm not saying that by any means, but I think it's a necessity for affordable housing.

Would you add anything to that, Ralph?
The only thing I might add and maybe Reed said this, but I think your partners in the development deal all along the tree have to have the same sort of mindset. I think more strategic, more specifically like the general contractor, right, like the general contractor has to have that same thought process.

It's just harder to absorb the litany of change orders that sometimes happen as it relates to general contractors. If you don't have a strategic partner that believes the same thing that you believe that is willing to take it on the chin on some small stuff, obviously you've got to if it if it's a big issue, you've got to address it. I mean, no hands down there. But, you know, in the world that I used to live in and specifically in some of the areas like Chicago and New York, a change order was off or the change order packages were often profit centers for general contractors as you went through the development process.

And I'm not saying that that doesn't happen in an affordable world, but if you have a community minded strategic partner in the construction field, those are a lot less likely to happen.

Great, great, thank you. Jeff, do you have anything you want to add? Yeah, it sounds like the state of South Carolina they're kind of doing a synthetic tiff. They are giving you back a portion of the taxes by, not by giving you a discounted rate. What is it that you have to do to trigger that C.R.? Is there a percentage of units that have to meet a percentage?

Well, let me back up a little bit. The county level is where it sits at. The ability to effectuate the tax is right, so with our SIRC at Robert Smalls, it was driven by a three-headed approach being the city, the county and the economic development entity, right? So, we were able to get the tax ceiling with a moderate growth per year for the development based on boxes that it checked. Obviously, one of those was also a commitment of X amount of dollars spent. It was discussions around minority and women-owned subcontractors. There was not a percentage that was put in front of us, but there was a commitment on our part to make best efforts to do that.

In other municipalities, there could be 20% of your overall budget goes to women or minority-owned contractors. That's not the case here, but that I've seen that happen before. And then you know that demolishing the eyesore that is Northridge within X amount of money, so it can be very deal-specific. But what I do say is that now post when we did this with Robert Smalls, the state themselves said, if you are doing affordable housing and you have a percentage of your time at 60 that you have, you have a non-profit partner in it that has a community focused 501(c)(3) designation .

You don't pay any taxes if you are meeting these criteria. And now it was through legislation passed in 2020 at the state level.

That's interesting how each state handles affordable housing and how they help by providing incentives. It's also interesting how they work or they don't. So, for the longest time, the state of South Carolina had no such, so we didn't even have a state match. So, the state, we have a federal four percent or the nine. We do not have state match until 2020. And we also didn't have the ability to have affordable taxes or affordable housing tax abatement based on a non-profit partner. So for the longest time and I think you saw maybe a drought, so to speak of affordable housing in the South Carolina Arena is that they did not have these live levers in place that neighboring places like North Carolina and Georgia had.

So, you can be a developer of affordable housing sitting in Spartanburg or Greenville and drive an hour and a half either way and be in states that have a four percent match and had the tax incentives to do so.

In Florida, we have entities that collect increment, you know, they do set a base here and anything that's derived over a period of time goes into a trust fund for the redevelopment agency. And then in 1980, there was a lawsuit that basically allowed these entities to bypass the state constitution and work directly with the private sector using those funds.

So, you know, we, you know, instead of giving a rebate on taxes, we actually take those dollars, which the state says are not necessarily taxes, and we're able to take your tax dollars that you pay and then in some fashion incentivize you. And again, you're right, it's at the local level who whichever entity has that agency, depending on what they want, they could say. For instance, in Fort Myers, they do minority-owned businesses. They also will give you money if you commit to make donations to a non-profit in their area to do things that they're not capable of doing or able to do. And it's just interesting how each state does it. And I was really interested and you know, the IRC and now this this this this nonprofit, do you think that with this nonprofit law that they just passed that, you know, the local governments looking at this and they're saying.

This might be a loaded question. You know, they have to make the do, is it the local government that makes the determination that if you have the non-profit, you get the tax rebate or the not even a tax rebate, but is it the state or is it is the local government, it's the State Department of revenue. OK, that makes sense because I could see some local governments who aren't really favorable to affordable housing, not giving that the tax deal because they want to keep the taxes. And but needless to say, it's interesting to see how each state does it.

Yeah, I agree. I agree. And you know, I think that's it for us. You know, we play in a lot of different locales and South Carolina is just kind of opened up the playbook a little bit. So, we're happy to be there.

One thing that you know again humbly, I do take pride in this, though. One thing that for the last 10 or 15 years, the Greeneville has done, because when you have a city that runs a surplus in a budget, you can have these.

Spartanburg is a little bit different for a lot of different reasons. You know, primarily we gave up the ability to annex in the 50s when we gave our water and sewer away. So, we can't really annex very well, like most cities can. So, we don't grow the way most people grow or historically didn't. So, we have a very artificially small city.

As an example. Durham has 300,000 people in the county. We have 300,000 people in our county. We have 40,000 people in our city. They have 260,000. So, it's just the ability to annex and grow is different. Long story short, they have been wanting to have an affordable housing trust fund to incentivize developers to do just what you're saying. Hey, listen, we have a toolbox. Here's some gap financing. Here's some X, Y and Z, whatever. Through the Robert Smalls development, we were able to seed that affordable housing trust fund for the first time and 10 or 15 years.

Because to your point, Stephen and Jeff, you've got to have these levers. You've got to have these tools and cities themselves have to be able to say, Hey, you know, Mr. Developer this is another way or carrot to help incentivize you to think about maybe putting in some affordable housing.

So, if you use the federal affordable housing, the nine in the four percent do you and this is just me not knowing in Florida, you have to hold that for 15 years. Yet, it has to remain to have that affordability for 15. How long would you project when you build these things that you know you're not a legacy builder, you're going to hold on to these things. How long do you think you would hold on to them?

So, after 15 years of wear and tear on a new project, it's going to take, you know, X dollars per unit to go in and rehab them. And it's kind of a mini restart on the whole deal. Again, so long as you're you have contracts and your deals with the, you know, state or federal contracts with HUD Align, you can rinse and repeat. That's kind of been our strategy is that, you know, at year 15, these are going to need a significant renovation redevelopment, and it's almost like doing another deal that you are done.

Guys, We are impressed, you know, it's hard to find developers who have a passion for community redevelopment that, you know, that also don't have a passion for a nice hefty profit margin. So, our hat's off to you for that.

Conclusion

14 · What Is the Future for CRED, and How You Can Utilize CRED-Oriented Ideas to Achieve Community-Focused Developments

Stephen Buckman, Jeff Burton, and John Talmage

Overview of the Book

The chapters in this book were designed to give you a broad overview of what it means to do community real estate development and how it differs from normal market-rate real estate development. We hoped that you viewed this book as just an overview and will now look deeper into more specific areas of CRED. The issues covered in the preceding 13 chapters of this book covered broad ideas of historical and theoretical backgrounds to the day in the life of the CRED developer.

In the previous 13 chapters, we gave you a broad umbrella of ideas, concepts, and stories to think about as you delve into the world of CRED. We began by discussing the historical background of how we got to this point in our urban and development policy. Chapter 2 discussed the foundational background of community real estate and community development policy, showing how the ideas around CRED in this book do not live in a vacuum. Instead, a long line of national and local policies has helped/attempted to cure urban society's social and design ills. On the backs of these past policies, we find ourselves in a world that is leaning more and more on the private development community with the help of the public sector, often in the form of public-private partnerships better known as P3, to enable the private sector to take the lead.

Moving on from a broad perspective of historical ways that policy has shaped the functions of CRED, the following five Chapters 3–7 looked at the mechanisms that can be implemented to help CRED-oriented developments move forward. Chapters 3 and 7 looked closely at the government-sponsored policies that help spur the private sector to take on developments that on their own would not pencil out. Chapter 3 covered some of the most commonly used and discussed ways (LIHTC, NMTC, and OZ) for the public sector to help subsidize the private sector. For instance, Low-Income Housing Tax Credits (LIHTC) have been arguably the most commonly used tax equity generating policy for developers to do affordable housing, while New Market Tax Credits (NMTC) and the newly devised Opportunity Zone

DOI: 10.1201/9781003109679-17

(OZ) legislation are designed to help spur neighborhood economic development through real estate development. The last significant government-sponsored tax structured policy is Tax Increment Financing (TIF). While LIHTC, NMTC, and OZ are all driven by the developer and around receiving of equity for the project, as Dr. Burton showed, TIF is a P3 with a municipality, based on issues such as blight as a requirement. However, the reliance on blight is becoming less critical as municipalities shift toward synthetic TIFs.

While Chapters 3 and 7 covered specific developer and state/local specific tax-driven direct support policies to encourage development, Chapters 4–6 take a more community-wide approach to the issue, ranging from historic preservation and the importance of historical buildings for community character. These chapters discuss the nuts and bolts of proper urban design. These chapters also highlight where and in what form CRED should adhere. Dr. Linkous particularly touched upon this topic in her chapter on Transfer of Development Rights (TDR) which openly discussed working with municipalities to find the proper site for one's development that fits within the community's overall design.

CRED and the policies that help produce community-oriented developments should be looked at positively and applauded, especially concerning straight market-driven development. But with anything good, there is also a dark side needing consideration; the dark side of CRED can result in gentrification. In Chapter 8, Drs. Bosman and Buckman discussed the resulting gentrification from these types of developments, emphasizing the viewpoints around class, race, and the real estate finance industry. The impact of gentrification should be seen as a cautionary tale as CRED is meant to produce good communities, homes, and economic benefits. Yet, rising home prices and land valuation may follow and become a genuine concern. CRED-oriented developments usually take place in communities that need an uplift. If developments of this nature succeed, the result may promote a community buzz and higher profile, invariably resulting in outside developers filtering into the area looking for cheap property in up-and-coming areas. Thus, just as uplifting the community is essential to its future prosperity, it is equally crucial for the CRED developer and local government to ensure that its existing citizens, who helped in the redevelopment process, can participate in its prosperity.

Following the CRED background discussion, which laid out its foundation, the book also talked with practitioners from the field. Through one-on-one discussions with multiple players, the reader understood how these fundamental processes played out on the ground. The voices from the field ran the gamut of lending perspectives (Joe Bonora) to realtors (Lisa Reaves) and housing authorities (Leroy Moore). In a sense, what should be taken from these viewpoints was how the background information discussed in Chapters 3–7 played out in practice. For instance, Reed Jones and Ralph Settle discussed using subsidies such as LIHTC to develop the Robert Smalls affordable housing project in Spartanburg, SC.

Why CRED Is Important

This book emphasizes the importance and needs for real estate development that takes into account more than just the return on investment (ROI) or the internal rate of return (IRR), but rather also takes into consideration how developers can serve the community and still profit by it. Also, the last few years, especially during the COVID-19 pandemic, have shown that CRED-driven ideas are needed more than ever.

Since the outbreak of COVID-19 in the United States in late 2019 and early 2020, we have seen that the urban realm impacted in two meaningful ways. First and foremost, we have seen the cost, whether it be for-sale housing or rental housing, skyrocket. This impact is especially the case and most shocking for places that have often been considered reasonable and affordable. For instance, the rental prices in Greenville, South Carolina (one of the editor's home cities) in one year (November 2020–November 2021) jumped 15%, making it cheaper to buy than rent even though the city also, as of January 2022, had one of the most overly inflated for-sale housing markets in the country (Stebbins, 2022). The various factors for this run from artificially suppressing interest rates by the Fed to remote workers now able to choose to live wherever they wished.

The results of all of this are the worsening of the already pre-existing condition prevalent in our society, with the valuation of profit over community resulting in housing insecurity for many. As was shown in the book, especially in Chapter 2, the United States has historically been plagued with housing insecurity issues. What COVID and rising rental rates and the unaffordability of housing ownership for many have shown us is that there is a considerable gap between those who need housing and how much housing is available. According to the National Low-Income Housing Coalition in the United States, there is a need for 6.8 million affordable housing units, and now in January 2022, there is no state in the country where someone making minimum wage and working full time can afford a two-bedroom apartment (nlihc.org).

The second significant impact on the built environment from COVID-19 is the push to work from home, and the reconfiguration of our work lives from office to remote work. With the increase in working from home, there has been a re-evaluation of one's community. People are now seeing their community in a new light and are looking for the community to supply infrastructure that was often taken care of outside of the community, such as child care, parks, schooling, after-school programs, and more. As citizens are rediscovering and reinvesting in their local communities, the need arises for developers to create developments that answer those needs.

CRED developers are uniquely positioned to answer those needs and fold them into their developments. Whether using LIHTC to bring affordable housing or proper urban design and other assets such as the e-gaming club for neighborhood kids (see Paramore e-gaming development in the appendix). These demands made by the community are to stay, and developers will need to develop accordingly.

K nowing that community and the reliance on the built form to ensure and help with society's social ills is here to stay, future developers must now more than ever learn the skills of what Community Real Estate Development presents. Through the CRED training that editors lead and a book like this, that has come out of that training, developers can begin to think about and learn about how to do community-driven real estate development. The onus will be on the developer, whether standard market rate, affordable housing, or the Housing Authority developer, to understand more about CRED because doing so will enable them to do good for the community and make money doing so. In the end, it can be a win-win for all parties involved if it is done correctly, which the CRED training and this book encourage.

Works Cited

NLIHC (2022). *Housing Needs By State*. www.nlihc.org. (retrieved 1.28.22).
Stebbins, S. (2022). "Greenville, SC Is One of the Most Overpriced Housing Markets in America." In *The Center Square*. www.thecentersquare.com (retrieved 1.28.22).

Overview of 2018 and 2019 Team Winners

Introduction to Examples of Class Winners

As touched upon in the introduction, the CRED program culminates in a competition between the students to determine which proposal is most fitting for a development. At the beginning of the six weekend class, development ideas are suggested and voted upon as to what the students want to work on. In general, four groups composed of five students per group move forward. Throughout the class, the instructors teach from a combination of lectures, guest lectures, and a hands-on approach using the teams' project as learning in action. The teams put together an entire development proposal which includes a market analysis, site analysis, proforma, and overall marketing sales package.

The culmination of this work comes on the last weekend when the teams put together their entire package into a written development proposal and present that proposal. The presentation takes place in front of a board of judges that runs the gamut from the instructors, guest speakers, as well as banking, development, and equity professionals from the community. The final proposals are then voted on in combination with the written proposal and presentation to determine which project is the most developable and the one that most covers the true spirit of CRED.

The winning proposal then is encouraged to move forward to make it a reality. Many of the judges on the panel do more than just encourage the winning team to move forward but help them in the process by helping with permitting, land acquisition, funding, etc. Furthermore, many of the ideas by students at the beginning of class are viable ideas with some backing before they even enter the world of CRED. One of the key important aspects of CRED is that many of the students do bring viable ideas to the class which they wish to learn how to make their ideas into a development reality. While the winners are of course encouraged to move forward, so are all the development proposals and often these do turn into real developments. This is truly one of the things that makes the CRED program great and that it is not just an academic exercise for many of the students.

What follows are two examples of "winning" proposals from the 2018 and 2019 CRED classes. As you can see, both proposals are very different in scope but each in their own way are examples of the importance of

DOI: 10.1201/9781003109679-18

community-oriented real estate development and the diversity in what can be done under the label of CRED. The first proposal comes from the 2018 class and is centered around an e-gaming arena in the Paramore District of Orlando. In this proposal, there is an e-gaming arena and condo hotel that helps to subsidize e-gaming community center for local youth; furthermore, the project would be loosely tied to the University of Central Florida's downtown campus which has one of the premier e-gaming degree programs in North America. While the 2018 winner looked to support community youth, the 2019 winner took a much more traditional approach to community real estate development by putting together a community land trust development concept that help spur homeownership in a disenfranchised area of Orlando. The project not only would help spur homeownership and create wealth for individuals but would also help to bring up the community.

Thus, the following two examples show the breadth of what CRED can be. As different as the two development proposals are, they are also highly similar. Both proposals in their own way are looking to help communities and individuals that make up those communities through real estate development.

Mixed-Use E-Sport Arena Development

Soomin Kim

Author Bio

Soomin Kim received a BA in History from Chung-Ang University in Seoul, Korea (2014). During his military service in United States Forces Korea, Combined Forces Command (USFK CFC), he did historic preservation research of US military bases in Korea (2011). He earned a MA in Urban Planning from the University of South Florida (2018). He worked as an intern at the Palmetto CRA, FL and participated in several community-based projects which gave him a great interest in the vulnerable senior population. His current interests are built in environmental history and senior health issues in city.

Project Team Members

The project team leader, Derek Watford, is co-founder of High-Point Gamer Co. and an E-sport consultant. Watford was the first developer for this project and has supported several related gaming/community groups. Dominique Cobb is the project's outreach specialist with vast experience in youth education. Cobb also contributed deep statistical and community analysis. Tyrone Walker is the program consultant for the project. His extensive experience in planning local educational programs and advice helped guide the project. Lastly, Soomin Kim is a community design specialist. His expertise was invaluable for exterior/interior design of facility.

Introduction

This chapter reports on a conceptual planning project aimed at educational gaming opportunities for the youth in Parramore, a west-central neighborhood of Orlando. Since computer games hit the market in 2007, gaming has consistently grown by almost 7.2% per year. The city of Orlando, Florida, is a leading location for this progressive industry. With year-over-year growth in tech jobs at 149%, a median population age of 37, and a unique community for independent game developers, it is on the verge of explosive growth

DOI: 10.1201/9781003109679-19

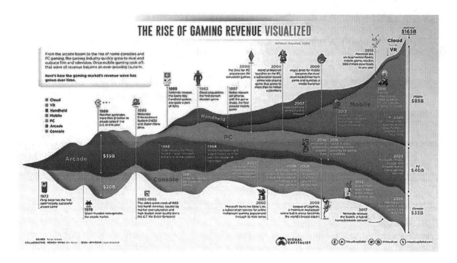

Figure 01.1
https://www.visualcapitalist.com/50-years-gaming-history-revenue-stream/.

in its game technology sector[1] (Figure 01.1)[2] and is No. 1 Metro with the Most STEM Job Growth[3] in 2018 and No. 5 Best City for Gamers[4] in 2021.

Development Overview

The team sought a project that would benefit both the local community and economic development. We believe private projects can produce educational benefits while contributing to vulnerable low-income and low-education communities. Thus, we agreed that youth should be our main target population.

We chose a mixed-use planning model to bring educational, revenue, and housing opportunities to Parramore. Orlando's significant interests and assets in the E-gaming industry made technology education a logical choice. Thus, we planned a year-round 15,000 sq. ft. E-sport arena as our primary facility. In addition, we included 15,000 sq. ft. of commercial/office space to provide flexibility in economic opportunities. Finally, 32 residential units would offer affordable housing and revenue generation. This mid-rise complex includes on-site structured parking (Figure 01.2).[5]

Projected Scenario (Proforma)

Hard projects costs are $21,506,020 and soft costs are $3,911,160 for a total of $25,417,180. At the beginning of our project, we received a $19,765,000 cash investment from a developer and Debt Financing Needed of $7,467,370. The financing rate was 6%. This funding brought our initial total to $12,297,630. Thanks to community interest and support, we received additional equity

Figure 01.2

Image depicts conceptual approach. Retrieved from https://www.bizjournals.com/sanfrancisco/news/2017/09/07/south-san-francisco-housing-summer-hill.html.

funding of $10,066,872, grant funding of $100,000, and loans of $15,250,308. The loan to value rate is 60%.

For project revenue, the 32 residential units, consisting of 16 one-bedroom/one-bathroom and 16 two-bedroom/two-bathroom units, will generate approximately $499,200 annually. We set the one-bedroom unit's monthly rent at $1,000 and two-bedroom units at $1,500 to ensure affordability for students. These rent rates will generate monthly revenue of $17,600 and $24,000, respectively. We project that arena rental will contribute $28,525 per month and office space rental $18,200. Additionally, we calculate annual naming rights income at $1,000,000 per year. In sum, total revenues will be $2,059,900 per year.

Existing Concept

There are three major E-gaming stadiums in the United States—Oakland, CA, Las Vegas, NV, and Los Angeles, CA. The city of Oakland supports numerous mixed-use planning and sports stadium redevelopment. For instance, the city constructed an E-gaming arena for 1,000 gamers as part of its Jack London Square commercial planning (Figure 01.3). The goal was to rehabilitate the local economy. On average, the existing baseball stadium held four minor leagues per week[6] and at least one major league per season. The Oakland arena demonstrates the attractiveness of E-sport arenas for millennials and commercial revenue.

The Luxor Hotel & Casino in Las Vegas, NV, is another creative example of mixed-use planning that incorporates E-gaming. At 30,000 square feet, the HyperX arena is one of the largest arenas nationwide (Figure 01.4). In the technical specifications, HyperX provides top-notch level gamer-friendly

Figure 01.3

https://www.bizjournals.com/sanfrancisco/news/2016/08/11/e-sports-gaming-arena-jack-london-square-oakland.html.

Figure 01.4

https://www.hyperxgaming.com/br/company/press/article/55345?ArticleTitle=HyperX%20and%20Allied%20Esports%20Renew%20Naming%20Rights%20Agreement%20for%20%20HyperX%20Esports%20Arena%20Las%20Vegas%20%20-%20(January%2013).

design while meeting the primary needs of casino visitors. Offerings include game-inspired food menus, E-sports Exhibition Show Matches, a vintage video game cocktail bar, 50-foot LED video wall, VIP Rooms, a luxury box suite lounge, broadcast center, and an independent production studio.

Luxor also supports gaming education. They provide education programs through their partner, Discord INC., one of the most prominent voice messenger companies worldwide. Notably, social equity is a priority for Discord. For instance, the 'Broadcast HER Academy' program encourages young women to enter the E-gaming broadcast sector.

Figure 01.5
https://www.musicgateway.com/blog/gaming-industry/games-business/best-gaming-conventions.

One of the most outstanding gaming companies worldwide, Blizzard, built an E-sports arena in Los Angeles, CA. Because the company owns the facility, it is a showplace for new technology presentations and educational sessions (Figure 01.5). It has also become one of the most iconic places for E-gaming enthusiasts and millennials who want to enter the profession. In addition, the arena is a tourist attraction for Burbank. In many ways, the arena has revived the region.

Market Overview

Orlando, FL, which is one hour away from Tampa, is growing significantly every year.[7] Its population's median age is 33. The city greatly supports sports entertainment and is home to world-class industry-leading game developers such as Activision Publishing. Additionally, the University of Central Florida provides proactive school program for video-game studies.[8] The city also hosts significant numbers of world tourists thanks to attractions such as Universal Studios and Disneyland.

Market Location

The New Orlando Development plan, The Creative Village, was the inspiration for our Parramore project. Creative Village contains the MLS Soccer Stadium, George C. Young Building, Sports Entertainment District, Amway Center, Police Headquarters, and the University of Central Florida (UCF) Campus. Furthermore, it targets authentic community life around downtown Orlando by ensuring a 15-minute walking radius (Figure 01.6). We believe the mixed-use E-sport arena is an excellent addition to this progressive community and its educational focus.

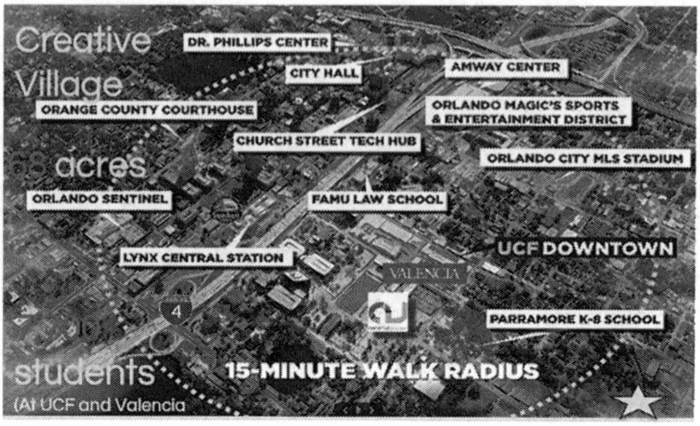

Figure 01.6
(Image edited by CRED team members).

The Community

Our target community was the Parramore neighborhood of Orlando. It is a historic African American neighborhood located west of downtown and close to the New Orlando Development project. There are 3,714 residents with a median household income of $17,618. Unfortunately, most residents do not have a college education, and median income levels are more than 50% lower than Orlando's median.[9] Our team felt that the project could bolster this community by strengthening local technology-based commercial and education opportunities.

Development Site

The site location chosen is 1212 W Livingston Street (Figure 01.7). This parcel is well suited for a community activity center. Required rezoning would be AC-1.[10]

Community Engagement

We recognized that community engagement was crucial for successful planning. Including the community in decision-making assures the development meets community needs. Specific areas identified health screenings, mentoring, visual arts experiences, league development, summer camps, and gaming-centric career introduction fairs, among others. Thus, we sought community partnerships to support the local economy.

Community Partners

We found many community organizations and groups willing to support our project. For instance, Orlando Magic is a local basketball team with significant interest in redefining the Parramore community. Many staff members

Figure 01.7
(Screenshot from GoogleMap, edited by MS Paint).

grew up in this area and showed their support and love for the community. A significant player in the gaming industry, Electronic Arts (EA), also committed to the project. Our team leader Derek also shared many resources from his start-up consulting company, High-Point Gamer. We also engaged UCF, Valencia College, Orlando Economic Partnership, and the city's planning and development entities.

For educational aspects, we reached out to the Wake Up Mentoring, Wells' Built Museum of African American History and Culture, and Parramore Kidz Zone (PKZ). They agreed with our educational focus on technology and E-gaming. They also encouraged including virtual reality (VR) technology education and training. To accommodate these activities, we included open space and interior designs to ensure adequate space and safety.

PKZ Program

Parramore has been Orlando's highest poverty neighborhood for more than a decade. As a result, the city launched the PKZ program to improve conditions for neighborhood youth (Figure 01.8) and committed to investing approximately $2.5 million a year. The goal was to build family stability and enhance academic success by increasing the number of youth entering and completing post-secondary education and keeping older youth out of trouble and on track toward social and economic success.[11] UCF is also heavily involved with the PKZ program. The team believed our mixed-use facility complemented both PKZ's and UCF's gaming educational efforts.

VR Technology

VR is currently one of the most novel technologies on the market. While VR technology has been around since the 1960s, it is gaining a stronger foothold

Figure 01.8
https://www.orlando.gov/Our-Government/Departments-Offices/FPR/Orlando-Kidz-Zones/Parramore-Kidz-Zone.

Figure 01.9
http://www.markmcguinn.com/virtual-reality-and-its-influence-on-content/.

due to improved hardware and stand-alone equipment (Figure 01.9). VR technology works with headgear and hand devices, creating an incredibly realistic 3D experience. Furthermore, many fields, such as real estate, medicine, excursions, etc., are applying this technology to provide immersive experiences and learning opportunities.

Our team is very interested in providing this new technology to the Parramore community. Because of its forecasted significance, it offers youth more opportunities in the future job market. Thus, we made sure to include a 'testing lab' in the commercial space. This versatile space allows for office and VR educational space.

Residential

Rental housing is targeted at the local young population pursuing gaming education. We expect that the complex will house over 7,000 students. We recognize that many college students are comfortable sharing multi-bedroom living quarters while others prefer single occupancy. Thus, we included two floor plans—two bedroom/two bath and one-bedroom/one bath units. Two-bedroom units average 1,500 and 800 sq. ft. for one-bedroom.

Commercial

EA Sports has a five-year lease for 3,000 sq. ft. commercial space, housing their QA and Testing Lab. The expected annual revenue is $218,400. Although not included in our initial revenue calculations, the complex also supports local retailers such as food stands or souvenir stores. Together with E-sports league merchandise and gaming gear stores, these offerings provide additional jobs for the community.

The E-sports arena is the primary design feature for the entire facility. The arena space occupies 15,000 sq. ft., which is average for gaming leagues. It will have a 500-seat capacity and live streaming capabilities. We anticipate yearly revenue of $1,342,300. Several considerations for this space included ensuring that the check-in area is large enough and the gaming space is flexible enough to accommodate seating for individual players and team configurations. This likely means adding risers and stages to the layout and building custom seating for players.

We also considered abundant seating for spectators and seating layouts that prioritize unobstructed views. For example, auditorium and theater seating are ideal layouts for esports as arenas need adequate space for massive LED screens and digital leaderboards. We also considered break-time areas and atmosphere such as lounges, snack stands, background music booths for DJ, bandstands, etc.[12]

Conclusion

The ongoing Covid-19 pandemic is expanding the gaming market. Stay-at-home orders, lockdowns, and social distancing have increased indoor activity and generated more attention to E-gaming leagues and product sales. This phenomenon significantly impacted younger generations such as Gen Z and Millennials, who spent more time gaming.[13] This increased interest offers significant opportunities for building on this momentum. E-gaming play and education facilities are among the mechanisms that communities can leverage.

Our project, Orlando Lobby, sought to serve a low-income, under-resourced community with a new wave of E-gaming and STEM technology education. Our team believes that mixed-use, community-based planning is critical for successful community development. We hope our plan can be a model for community engagement and planning that embraces and supports economic growth, youth, and future career pathways.

Works Cited

City of Orlando, (2021) *Parramore Kidz Zone*. Retrieved from https://www.orlando.gov/Our-Government/Departments-Offices/FPR/Orlando-Kidz-Zones/Parramore-Kidz-Zone

Clement, J., (2021, January 4). *COVID-19 Impact on the Gaming Industry Worldwide - Statistics & Facts*. Retrieved from https://www.statista.com/topics/8016/covid-19-impact-on-the-gaming-industry-worldwide/

Coliseum, (2018, August 14). *This Chinese Stadium Is Exclusively Dedicated to Esports*. Retrieved from https://www.coliseum-online.com/exclusively-e-sports-chinese-stadium/

Keller, M., (2019, March 27). *Serious About Play, Orlando Becomes Leader In Multibillion-Dollar Video Game Industry*. Retrieved from https://www.forbes.com/sites/orlando/2019/03/27/serious-about-play-orlando-becomes-leader-in-multibillion-dollar-video-game-industry/?sh=46cbb9a13b5a, https://www.city-data.com/city/Orlando-Florida.html

Kotkin, J., (2018, January 11). *Tech's New Hotbeds: Cities With Fastest Growth In STEM Jobs Are Far From Silicon Valley*. Retrieved from https://www.forbes.com/sites/joelkotkin/2018/01/11/techs-new-hotbeds-cities-with-fastest-growth-in-stem-jobs-are-far-from-silicon-valley/?sh=5a8aada4bed1

Reyes-Velarde, A., (2017, September 07). *Housing Project May Help South San Francisco*. Retrieved from https://www.bizjournals.com/sanfrancisco/news/2017/09/07/south-san-francisco-housing-summer-hill.html

Roche, A., (2018, January 11). *Orlando's Fast-Growing Game Technology Sector*. Retrieved from https://news.orlando.org/blog/orlandos-fast-growing-game-technology-sector/

Social Tables, *Esports Venue Requirements: The Essential Guide*. Retrieved from https://www.socialtables.com/blog/event-planning/esports-venue-requirements/

Stone, C., (2020, January 24). *Orlando Green Lights Magic's Sports and Entertainment District Amendments* Retrieved from https://www.mynews13.com/fl/orlando/news/2020/01/24/orlando-green-lights-magics-sports-and-entertainment-district-amendments

Wallach, O.,(2020, November 23). *50 Years of Gaming History, by Revenue Stream (1970–2020)*. Retrieved from https://www.visualcapitalist.com/50-years-gaming-history-revenue-stream/

Notes

1 https://news.orlando.org/blog/orlandos-fast-growing-game-technology-sector/.
2 https://www.visualcapitalist.com/50-years-gaming-history-revenue-stream/.
3 https://www.forbes.com/sites/joelkotkin/2018/01/11/techs-new-hotbeds-cities-with-fastest-growth-in-stem-jobs-are-far-from-silicon-valley/?sh=5a8aada4bed1.
4 https://wallethub.com/edu/best-cities-for-gamers/36270.

5 Image depicts conceptual approach. Retrieved from https://www.bizjournals. com/sanfrancisco/news/2017/09/07/south-san-francisco-housing-summer-hill. html.

6 Before COVID pandemic.

7 https://www.mynews13.com/fl/orlando/news/2020/01/24/orlando-green-lights-magics-sports-and-entertainment-district-amendments.

8 https://www.forbes.com/sites/orlando/2019/03/27/serious-about-play-orlando-becomes-leader-in-multibillion-dollar-video-game-industry/?sh=6224484d3b5a.

9 https://www.city-data.com/neighborhood/Parramore-Orlando-FL.html.

10 Activity Center Zoning District.

11 https://www.orlando.gov/Our-Government/Departments-Offices/FPR/Orlando-Kidz-Zones/Parramore-Kidz-Zone.

12 https://www.socialtables.com/blog/event-planning/esports-venue-requirements/.

13 https://www.statista.com/topics/8016/covid-19-impact-on-the-gaming-industry-worldwide/.

Winner 2019

Orange Center Boulevard Redevelopment

Camille Reynolds Lewis

Author Bio

Camille is executive director for the Hannibal Square Community Land Trust (HSCLT) in Winter Park, FL. She served on the board of HSCLT for 5 years with 1 year as board president prior to being selected as its executive director. Camille's professional background is in public administration. She served as Environmental Programs Administrator for Orange County government and Senior Aide to the Chief Administrative Officer and Capital Projects Manager with the City of Orlando.

Project Overview

Through a competitive solicitation, the City of Orlando selected Hannibal Square Community Land Trust to redevelop six (6) multi-family residential parcels at the southwest corner of Orange Center Boulevard and Tampa Avenue in Orlando. Our proposal for the site is a mixed-use, mixed-income development—providing 30 townhomes for homeownership, 28 apartments, and 15,472 SF of community-based retail space.

This development will bring brand new homes for homeownership in an area where most of the existing homes were built between the 1950s and 1970s, and many houses are currently rentals with absentee landlords (Image 02.1).

This project will redevelop six (6) multi-family residential parcels at the southwest corner of Orange Center Boulevard and Tampa Avenue in Orlando, located in a designated opportunity zone in Census Tract 117.02. Our proposal for the site is a mixed-use, mixed-income development—providing 30 townhomes for homeownership, and a 3.5-story mixed-use building featuring 28 apartments, rooftop amenities, and 15,472 SF of retail space.

The surrounding neighborhoods are long-standing, close-knit, historically African-American, middle-class neighborhoods; however, blight and vacant housing have been an ongoing issue since the late 1990s. While the single-family and multi-family housing stock is good comprising mostly concrete block units, most of the existing homes were built between the 1950s and 1970s, and many houses are currently rentals with absentee landlords.

DOI: 10.1201/9781003109679-20

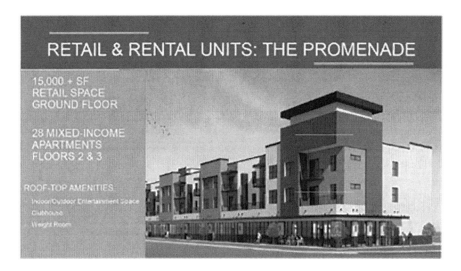

Image 02.1

Most of the current homeowners are long-standing residents, having lived there for generations. There is retail in the commercially-zoned areas of the community, but much of it is the retail typically associated with blighted neighborhoods, having deteriorating buildings and associated parking lots.

Homeownership and Long-Term Affordability

Roughly 32% of the homes in the area are owner-occupied while more than 48% of the population is renting. This development will bring 30 brand new homes for homeownership. That is roughly a 5% increase in the homeownership rate. The proposed development will help improve property values while increasing the number of owner-occupied units in the area.

Research has proven that, in addition to being instrumental in narrowing the wealth gap, affordable homeownership stabilizes neighborhoods and boosts resident engagement. Homeownership encourages property upkeep, which leads to higher home values. In addition, homeowners tend to be more civically engaged and volunteer in their communities at higher rates than renters.

Site Details

The site consists of 4.68 acres that previously housed six buildings with a total of 111 units of low-income apartments that were built in the 1950s. Over the decades, the apartment buildings went into disrepair and foreclosure and were eventually acquired by the City of Orlando. The site is currently vacant (no one was displaced or is being displaced by the redevelopment). It sits directly across the street from the new Pendana apartments and senior living apartment building.

It is diagonally across from Tinker Field, in walking distance of Camping World Stadium, and directly north of George Barker Park on Clear Lake. The site is also a two-minute drive to the East/West Expressway (408) and less than 10 minutes from I-4 and downtown Orlando, the Amway Center, Valencia Downtown, and Exploria Stadium—home of the Orlando Pride soccer team. The area is in the heart of Orlando with easy access to shopping, schools, and entertainment. Lynx public transportation is also within comfortable walking distance.

Geographic Area

The proposed development is in designated opportunity zone (Census Tract 117.02) in the West Lakes neighborhoods, west of downtown Orlando.

Five neighborhoods and four lakes make up the West Lakes community. The neighborhoods are long-standing, close-knit, historically African-American, middle-class neighborhoods; however, blight and vacant housing have been an ongoing issue in the community since the late 1990s. At the same time, the single-family and multi-family housing stock is good stock, comprising many concrete block units. Many of the current homeowners are long-standing residents whose incomes range from stable to low and/or fixed.

Five Neighborhoods:

- Luola Terrance/Lake Sunset
- Lake Lorna Doone
- Rock Lake
- Clearlake Cove and Clearlake
- Bunche Manor/Hollando

Four Beautiful Lakes:

- Lake Sunset
- Lake Lorna Doone
- Rock Lake
- Clear Lake

Two Major Parks:

- George Barker Park
- Lizzie Rogers Park

Community and Neighborhood Activity Centers:

*   Jones High School
*   Camping World Stadium
*   Tinker Field

Townhomes for Homeownership

HSCLT will build 30 townhomes for homeownership on the parcels facing Orange Center Blvd. The townhomes are approximately 1,664 SF, 3BR/2.5BA, with detached, two-car garages, private courtyards, and master suite with private balcony. Construction material is a concrete block on both floors. Parking is to the rear of the units via private road.

The homes feature stainless steel appliances, solid surface countertops, wood-look, vinyl floor planks, spacious rooms, and an open floor plan.

Sales Set-asides by AMI:

*   15 units for families—65% to 80% of AMI
*   9 units for families—81% to 120% AMI
*   6 units at any income

Neighborhood Activity Center

The mixed-use portion of the property sits at the corner of Orange Center Boulevard and Tampa Avenue and directly in front of George Barker Park on Clear Lake. George Barker Park is a seven-acre neighborhood park, originally developed in 1968 and named for George A. Barker Jr., a former City Commissioner. It has three pavilions, a boat ramp, two fishing piers, playground, basketball court, and volleyball court. It sits on the northeast corner of Clear Lake and since Fall of 2020, it hosts a weekly farmers market. It also sits diagonally across the street from Tinker Field. Tinker Field was constructed in 1914 and served as the spring training home of the Brooklyn Dodgers, Cincinnati Reds, Washington Senators, and Minnesota Twins from 1923 through 1935. It was also the home park of the Orlando Rays minor league baseball team. In 1964, Dr. Martin Luther King, Jr. gave his only public speech in the city of Orlando from the pitcher's mound in Tinker Field.

This portion of the site will be developed after 75% of the townhomes are pre-sold and at least 70% have certificate of occupancy.

Phase II of the project is a 3.5-story, mixed-use development planned for 1.36 acres at the southwest corner of Orange Center Blvd and Tampa Avenue. The building will consist of 15,472 SF of retail space on the ground floor, 28 1- and 2-bedroom apartments (14 each on the 2nd and 3rd floors), a small community room, and rooftop amenities. The rooftop will have indoor and outdoor entertaining space that can be leased for private functions as well as a small clubhouse for the residents. The clubhouse is also available for use by residents of the townhomes as part of their HOA membership. The retail spaces on the ground floor will primarily be micro-retail with an anchor retailer, preferably a small, fresh produce grocer. The ground floor will also feature opportunity to host indoor/outdoor events that incorporate the small section of road between the development site and the City of Orlando's lift station. The City has said that there is opportunity to close that off for planned functions. Parking is in the rear of the building and on the small section of Tampa Avenue in front of the building.

Safe and Decent Rentals

The Pendana at West Lake apartments across the street from this site was built in 2018 by LIFT Orlando in partnership with Columbia Residential. They feature 206 mixed-income 1-, 2-, and 3-bedroom apartments with community amenities such as pocket parks, courtyards, and a pool.

Our proposed mixed-use development at the corner of Orange Center Boulevard and Tampa Avenue includes 28 apartments evenly spread across the 2nd and 3rd floors. The 1BR/1BA and 2BR/2BA apartments target mature adults looking for quality rental without the luxury price tag. The rooftop will have indoor and outdoor entertaining space that can be leased for private functions as well as a small clubhouse and weight room for the residents. The clubhouse is available for use by residents of the townhomes as part of their HOA dues. Our proximity to downtown Orlando, Interstate-4, the East-West Expressway (SR 408), and the local bus line, makes the apartments ideal for working adults traveling to work in any part of the Orlando area. Having access to the retail mix downstairs will provide an added convenience.

Retail Space and Jobs

Roughly 10,000 vehicles travel across Orange Center Boulevard daily, providing a source of traffic to the retail spaces that will anchor the development. In addition, the residents of the 200 apartments at Pendana, residents from the five surrounding neighborhoods, and residents of our planned townhomes and apartments will add foot-traffic for the retailers. The events at Tinker Field and Camping World Stadium, just two blocks north, can also provide an influx of shoppers.

The retail space will provide opportunity for community-based businesses to sell their goods and services in attractive space in a prime location. We envision this site as a potential economic engine for the surrounding neighborhood. We also anticipate that the site will create new, quality, decent-waged jobs for residents in the local area.

Much of the space will be micro-retail units for local vendors (such as small eateries, a satellite bank, or a barbershop). A fresh produce bodega and deli (the area is currently a "food desert") could anchor the site. The Black Business Investment Fund (BBIF) of Central Florida has expressed interest in

Image 02.2

providing technical and financial assistance to the retailers to help them be successful. These vendors could potentially hire from the surrounding neighborhood. For example, the bodega will need staff to stock inventory and serve as cashiers. The building also features community gathering/meeting space and an indoor/outdoor venue for private rentals and small, local events (such as a food truck mash-up, or event promoting local artists).

This locally focused development is revitalization without gentrification:

- Opportunity for small, local businesses.
- Community gathering space and events.
- Short-term benefit from construction jobs: Partnership with CTG; opportunity to learn a new trade for some living in the area.
- Long-term benefit from retail jobs: According to the U.S. Small Business Administration, small businesses generated 44% of the United States' economic activity from 1998 to 2014.

Mixed-Use Development Budget

The development budget for the mixed-use portion is $8.6M as of August 2021. The Black Economic Development Fund intends to use proceeds from the sales of the townhomes toward the equity requirement for the mixed-use. The details of this financing still need to be flushed out and construction costs will need to be rebid. See the link below for the current budget (Image 02.2).

Appendix
Glossary of Key CRED Terms

Adjustable-Rate Mortgage (ARM)   A mortgage loan with an interest rate is subject to change over the term of the loan. The interest rate is tied to the performance of a specified market rate.

Amortization   Amortization is the act of paying down principal over its term. In a typical mortgage loan, the principal is scheduled to be paid off or fully amortized over the loan term.

Average Hourly Earnings   A monthly reading by the Bureau of Labor Statistics of the earnings of hourly plant and non-supervisory workers in the private sector.

Basis Point   A basis point is one one-hundredth of a percentage point. For example, if mortgage rates fall from 7.50% to 7.47%, they have declined three basis points. A full percentage point is 100 basis points.

Brownfield   With specific legal exclusions and additions, the term "brownfield site" means real property, the expansion, redevelopment, or reuse, which may be complicated by the presence or potential presence of a hazardous substance, pollutant, or contaminant. Cleaning up and reinvesting in these properties protects the environment, reduces blight, and takes development pressures off greenspaces and working lands.

Cash-Out Refi   Refinancing a mortgage in which the new principal (the borrowed amount) exceeds the original loan's outstanding principal by at least 5%. In other words, the homeowner is taking equity out of the home.

Census Tract   An area delineated by the U.S. Bureau of the Census for which statistics are published; in urban areas, census tracts correspond roughly to neighborhoods.

Central Business District   Central Business District (CBD). Office buildings located in the central business district are in the heart of a city. In larger cities like Chicago or New York, and some medium-sized cities like Orlando or Jacksonville, these buildings would include highrises in downtown areas. Other office buildings can be found in other areas of the metropolitan area, including suburban office buildings. This office space classification generally includes mid-rise structures of

80,000–400,000 sq. ft. located outside of a city center. Cities will also often have suburban office parks that assemble several different mid-rise buildings into a campus-like setting.

Cityscape   A cityscape is the urban equivalent of a landscape.

Co-Housing Projects   Co-housing projects are the intentional community of private homes clustered around shared space. Each attached or single-family home has standard amenities, including a private kitchen. Shared space typically features a typical house, including a large kitchen and dining area, laundry, and recreational spaces. Shared outdoor space may include parking, walkways, open space, and gardens. Neighbors also share resources like tools and lawnmowers.

Commercial Development   Commercial property (also called commercial real estate, investment, or income property) refers to buildings or land intended to generate a profit, either from capital gain or rental income. Commercial property includes office buildings, industrial property, medical centers, hotels, malls, retail stores, farmland, multifamily housing buildings, warehouses, and garages. In many states, residential property containing more than a certain number of units qualifies as a commercial property for borrowing and tax purposes.

Commercialization   Commercialization is transforming an area of a city into a place attractive to residents and tourists alike in terms of economic activity.

Concurrency   "Concurrency" is a shorthand expression for a set of land use regulations that local governments are required (by the Florida Legislature) to adopt to ensure that new development does not outstrip local government's ability to handle it. For a real estate development to "be concurrent" or "meet concurrency," the local government must have enough infrastructure capacity to serve each proposed real estate development. Specifically, concurrency regulations require that the local government has stormwater, parks, solid waste, water, sewer, and mass transit facilities to serve each proposed development. Together, these seven public services are known as "concurrency facilities."

Conforming Mortgage Loan   Any mortgage loan at or below the amount Fannie Mae and Freddie Mac can purchase and/or securitize in the secondary mortgage market.

Construction Loan   A construction loan is a temporary loan used to pay for the building of a house.

Consumer Confidence Index   A measure of confidence households have in the economy. The index is released monthly by the Conference Board.

Consumer Price Index (CPI)   The CPI is the measurement of the average change in prices paid by consumers for a fixed market basket of a wide variety of goods and services to determine inflation's underlying rate. The broadest and most quoted CPI figure reflects the average change in the prices paid by urban consumers (about 80% of the U.S. population). The so-called "core CPI" excludes the volatile food and energy sectors.

Conventional Mortgage Loan   A conventional mortgage loan is any mortgage loan not guaranteed or insured by the government (typically through FHA or V.A. programs).

Credit Report   A credit report is an individual's report of borrowing and repayment history.

Credit Score   A credit score is a three-digit number based on an individual's credit report used to indicate credit risk.

Edge City   An edge city is a large node of office and retail active activities on an urban area's edge.

Employment (Payroll)   The number of non-farm employees on the payrolls of more than 500 private and public industries, issued monthly by the Bureau of Labor Statistics.

Employment Cost Index   An employment cost index is a quarterly index used to gauge the change in civilian labor cost, including salaried workers.

Existing Home Sales   Based on the number of closings during a particular month. Because of the one-to-two month period between a signed purchase contract and a closing, existing home sales are more influenced by mortgage rates a month or two earlier than the prevailing mortgage rate during the month of closing.

Fannie Mae and Freddie Mac   Fannie Mae and Freddie Mac are the nation's two federally chartered and stockholder-owned mortgage finance companies. Forbidden by their charters from originating loans (that is, from providing mortgage loans on a retail basis), these two Government-Sponsored Enterprises (GSEs) purchase and/or securitize mortgage loans made by others. Due to their directive to serve low-, moderate-, and middle-income families, the GSEs have loan limits on the purchase or securitization of mortgages.

Federal Funds Rate   The rate banks charge each other on overnight loans made between them. These loans are generally made so that banks can cover their daily cash flow and reserve requirements. The federal government doesn't actually set the fed funds rate, which is determined by the funds' supply and demand. Instead, it forms a target rate and affects the supply of funds through its securities sales purchases.

Federal Open Market Committee (FOMC)   The Federal Reserve arm that sets monetary policy, the FOMC, is scheduled to meet eight times a year. The 12 members of the FOMC include the seven governors of the Federal Reserve System, the president of the New York Federal Reserve Bank, and, on a rotating basis, four of the presidents from 11 other regional Federal Reserve Banks.

Fixed-Rate Mortgage (FRM)   A fixed-rate mortgage is a mortgage loan with an interest rate that does not change over the loan term.

Gentrification – This Is a Controversial Definition   Gentrification is converting an urban neighborhood from a predominantly low-income renter-occupied area to a mostly middle-class owner-occupied area. The renewal and rebuilding process accompanying the influx of middle-class or affluent people into deteriorating areas often displaces lower-income residents.

Greenbelt   A ring of land maintained as parks, agriculture, or other open space types limits the sprawl of an urban area.

Greenfield   A greenfield project is one that lacks constraints imposed by prior work. The analogy is to construction on greenfield land where there is no need to work within existing buildings or infrastructure constraints.

Gross Domestic Product (GDP)   The value of all the final goods and services produced in the United States over a particular period. Available quarterly from the Bureau of Economic Analysis.

Home Equity   Home equity is the difference between the house's current value and the amount of money owed on the mortgage.

Home Equity Line of Credit   A home equity line of credit is an open credit line secured by the equity in your home.

Home Improvement Loan   Money lent to a property owner for home repairs and remodeling.

Homeownership Rate   The homeownership rate is the number of households residing in their own home divided by the total number of households in the U.S. The U.S. Census Bureau releases an estimate of the homeownership rate based on a quarterly survey.

Hotels

Full-Service Hotels   Full-service hotels are usually located in central business districts or tourist areas and include the big-name flags like Four Seasons, Marriott, or Ritz Carlton.

Limited Service Hotels   Hotels in the limited service category are usually boutique properties. These hotels are smaller and don't typically provide amenities such as room service, on-site restaurants, or convention space.

Extended Stay Hotels   These hotels have larger rooms, small kitchens, and are designed for people staying a week or more.

House Price Index   A quarterly measure of the change in single-family house prices released by the Office of Federal Housing Enterprise Oversight. The HPI is a repeat sales index, meaning it measures average price changes in repeat sales or refinancing on the same properties. It is based on mortgages purchased or securitized by Fannie Mae and Freddie Mac. Homes with mortgages above the Fannie/Freddie conforming loan limit and houses insured or guaranteed by the FHA, VA, or other federal government entity are not included in the sampling.

Housing Starts   The Census Bureau's monthly count of the number of private residential structures on which construction has started or permits has been issued.

Industrial

Heavy Manufacturing   This category of industrial property is a particular use category for most large manufacturers. These properties are heavily customized with machinery for the end-user and usually require substantial renovation to re-purpose for another tenant.

Light Assembly These structures are much simpler than the above heavy manufacturing properties and usually can be easily reconfigured. Typical uses include storage, product assembly, and office space.

Flex Warehouse Flex space is industrial property that can be easily converted and typically includes a mix of industrial and office space.

Bulk Warehouse These properties are extensive, generally in the range of 50,000–1,000,000 sq. ft. These properties are often used for regional distribution of products and require easy access by trucks entering and exiting highway systems.

Infill Infill is building on empty parcels of land within a checkerboard pattern of development.

Informal Sector Economic activities that take place beyond official record, not subject to formalized systems of regulations or remuneration.

Infrastructure Infrastructure is the underlying framework of services and amenities needed to facilitate productive activity.

Interest Rate Interest rate is a measure of the cost of borrowing.

Jumbo Mortgage Loan A jumbo mortgage loan is a mortgage loan for an amount exceeding the Fannie Mae and Freddie Mac loan limit. Because the two agencies can't purchase the lender's loan, jumbo loans carry higher interest rates.

Lateral Commute Traveling from one suburb to another and going from home to work.

Lift Station Wastewater lift stations are facilities designed to move wastewater from lower to higher elevation, mainly where the source's height is not sufficient for gravity flow and/or when the use of gravity conveyance will result in excessive excavation depths and high sewer construction costs.

Loan-To-Value Ratio (LTV) In a mortgage loan, the amount borrowed relative to the value of the property. An LTV of 80% means the mortgage loan is 80% of the property's value, with the borrower making a 20% down payment.

Mean Home Price (of New or Existing Homes Sold) The mean home price is a mathematical average of the costs of all homes sold in the period, typically monthly. The mean price of homes sold generally runs higher than the median price due to the number of very high-priced homes.

Median Home Price (of New or Existing Homes Sold) The median home price is the median price of all the homes sold within 30 days. Median home prices are generally a better indicator of home price trends than average home prices.

Megacities Megacities are cities with more than 10 million people.

Megalopolis Megalopolis is the continuous urban complex in the Northeastern United States.

Metropolitan Area   Within the United States, an urban area consisting of one or more whole country units, usually containing several urbanized areas or suburbs, all act together as a coherent economic whole.

Metropolitan Statistical Area (MSA)   In the United States, an urbanized area of at least 50,000 population, the country within which the city is located and adjacent countries meeting one of several tests indicating a functional connection to the central city.

Micropolitan Statistical Area   A micropolitan statistical area is an urbanized area of between 10,000 and 5000 inhabitants, the country in which it is found, and adjacent countries tied to the city.

Mixed-Income   Mixed-income housing may include housing that is priced based on the dominant housing market (market-rate units), with only a few units priced for lower-income residents. It may not have any market-rate units and be built exclusively for low- and moderate-income residents.

Mixed-Use   In a broad sense, mixed-use development—any urban, suburban, or village development, or even a single building—blends a combination of residential, commercial, cultural, institutional, or industrial uses. Those functions are physically and functionally integrated, and that provides pedestrian connections.

Mortgage   A loan is lent to buy real estate and secured by the real estate.

Mortgage Application Index (Purchase)   An index published weekly by the Mortgage Bankers Association of America gauges the number of applications submitted to purchase a home. The survey covers about 40% of all retail, residential mortgage transactions.

Mortgage Application Index (Refinance)   A mortgage application index (refinance) is an index published weekly by the Mortgage Bankers Association of America gauges the number of applications submitted for a home's refinancing. The survey covers about 40% of all retail, residential mortgage transactions.

Mortgage Broker   A person or company that acts as a mediator between borrowers and lenders.

Multi-family Housing   Multifamily residential (also known as a multi-dwelling unit or MDU) is a classification of housing where multiple separate housing units for residential inhabitants are contained within one building or several buildings within one complex. A common form is an apartment building. Sometimes, units in a multifamily residential building are condominiums, where the units are typically owned individually rather than leased from a single apartment building owner. Many intentional communities incorporate multifamily residences, such as in co-housing projects.

Garden Apartments   Suburban garden apartments started popping up in the 1960s and 1970s, as families moved from urban centers to the suburbs. Garden apartments are typically 3–4 stories with 50–400 units, no elevators, and surface parking.

Mid-rise Apartments These properties are usually 5–9 stories, with between 30 and 110 apartment units and elevator service. These are often constructed in urban infill locations.

High-rise Apartments High-rise apartments are found in larger markets, usually have 100+ units, and are professionally managed.

Multiplier Effect The direct, indirect, and induced consequences of change in an activity; in urban geography, the expected addition of non-basic workers and dependents to a city's local employment and a population that accompanies new primary sector employment.

Office Classification

Office Buildings Are Usually Loosely Grouped into One of Three Categories: Class A, Class B, or Class C. These classifications are all relative and largely depend on context. Class A buildings are considered the best of the best in terms of construction and location.

Class B properties might have high-quality construction but with a less desirable location.

Class C is everything else.

Office Park An office park is an area where several office buildings are built together on landscaped grounds.

Planned Communities A planned community is a city, town, or community designed from scratch and developed to follow the plan.

Program-Related Investment Program-related investments (PRIs) hold incredible potential for the social enterprise arena. Rather than giving away money through grants, PRIs allow foundations to make investments as loans or equity stakes in the hopes of regaining their assets plus a reasonable rate of return.

Producer Price Index (PPI) The PPI measures the average change in the selling prices of goods and services sold by domestic producers and an inflation indicator and released monthly by the Bureau of Labor Statistics.

Racial Steering The practice in which real estate brokers guide prospective home buyers toward or away from specific neighborhoods based on their race.

Redlining A process by which banks draw lines on a map and refuse to let money to purchase or improve the property within the boundaries.

Real Estate Owned REO is a term used in the United States to describe a class of property owned by a lender—typically a bank, government agency, or government loan insurer—after an unsuccessful sale at a foreclosure auction.

Restrictive Covenants Restrictive covenants are statements written into a property deed that restricts the use of land in some way.

Retail

Strip Center Strip centers are smaller retail properties that may or may not contain anchor tenants. An anchor tenant is simply a larger retail tenant, which usually draws customers into the property. Examples of anchor tenants are Wal-Mart, Publix, or Home Depot. Strip centers typically contain a mix of small retail stores like Chinese restaurants, dry cleaners, nail salons, etc.

Community Retail Center Community retail centers are generally in the range of 150,000–350,000 sq. ft. Multiple anchors occupy community centers, such as grocery stores and drug stores. Additionally, it is common to find one or more restaurants located in a community retail center.

Power Center A power center generally has several smaller, inline retail stores but is distinguished by a few major box retailers, such as Wal-Mart, Lowes, Staples, Best Buy, etc. Each big-box retailer usually occupies between 30,000–200,000 sq. ft., and these retail centers typically contain several out parcels.

Regional Mall Malls range from 400,000–2,000,000 sq. ft. and generally have a handful of anchor tenants such as department stores or big-box retailers like Barnes & Noble or Best Buy.

Out Parcel Most larger retail centers contain one or more out parcels, which are parcels of land set aside for individual tenants such as fast-food restaurants or banks.

Retention/Detention Pond A detention pond is a low-lying area designed to temporarily hold a set amount of water while slowly draining to another location. These ponds control floods when large amounts of rain could cause flash flooding. A retention pond is designed to hold a specific amount of water indefinitely.

Second Mortgage A second mortgage is a real estate mortgage that has already been pledged as collateral against another mortgage. Typically used to draw cash from home for other purposes.

Securitization Securitization is the pooling of mortgage loans into a mortgage-backed security. The principal and interest payments from the individual mortgages are paid out to the MBS security holders.

Short Sale A short sale is a sale of real estate in which the net proceeds from selling the property will fall short of the debts secured by liens against the property. In this case, if all lien holders agree to accept less than the amount owed on the debt, a sale of the property can be accomplished.

Single-family Housing A single-family (home, house, or dwelling) means that the building is usually occupied by just one household or family and consists of only one dwelling unit or suite. In some jurisdictions, allowances are made for basement suites or mother-in-law suites without changing the "single family" description. However, it does exclude any short-term accommodation (hotel, motels, inns) and large-scale rental accommodation (rooming or boarding houses, apartments).

Smart Growth Smart growth is legislation and regulations to limit suburban sprawl and preserve farmland.

Sprawl Sprawl is developing new housing sites with relatively low density at locations that are not contiguous to the existing built-up area.

Suburbanization Suburbanization is the movement of upper and middle-class people from urban core areas to the surrounding outskirts to escape pollution as well as deteriorating social conditions.

Topography The topography is a detailed map of the surface features of the land. It includes the mountains, hills, creeks, and other bumps and lumps on a particular hunk of earth.

Underwriting The determination of the risk a lender would assume if a particular mortgage loan application is approved.

Urban Growth Rate The urban growth rate is the rate of growth of an urban population.

Urban Morphology The form and structure of cities, including street patterns and the size and shape of buildings.

Urbanized area—the city as art.

Zoning Dividing an area into zones or sections reserved for different purposes such as residence and business and manufacturing.

Index

Printed in the United States
by Baker & Taylor Publisher Services